Other books by Gerald Feinberg:

Life Beyond Earth by Gerald Feinberg and Robert Shapiro
Consequences of Growth
What Is the World Made Of?
The Prometheus Project

Solid Clues

Quantum Physics, Molecular Biology, and the Future of Science

by Gerald Feinberg

A TOUCHSTONE BOOK
Published by Simon & Schuster, Inc.
NEW YORK

First Touchstone Edition, 1986

Published by Simon & Schuster, Inc.
Simon & Schuster Building
Rockefeller Center
1230 Avenue of the Americas
New York, New York 10020

TOUCHSTONE and colophon are registered trademarks
of Simon & Schuster, Inc.

Designed by Barbara Marks

Manufactured in the United States of America

10 9 8 7 6 5 4 3 2 1
10 9 8 7 6 5 4 3 2 1 Pbk.

Library of Congress Cataloging in Publication Data
Feinberg, Gerald, date.
   Solid clues.

   Bibliography: p.
   Includes index.
   1. Science. 2. Physics. 3. Molecular biology.
4. Quantum theory.   I. Title.
Q158.5.F45   1985   500   84-27689
ISBN: 0-671-45608-3
ISBN: 0-671-62252-8 Pbk.

Acknowledgments

A book such as this could not have been written without substantial assistance from many people, and I am pleased to be able to thank those who have given me that assistance. Many scientists have generously discussed with me their views about the current state and future prospects of their own branch of science. In physics, these included Drs. Norman Christ, Sheldon Glashow, Tsung-dao Lee, I. I. Rabi, Richard Osgood, Edward Spiegel, and Erick Weinberg; in chemistry, Drs. Philip Pechukas, and Robert Shapiro; in biology, Peter Gouras, Drs. John Harding, Cyrus Levinthal, and Robert Pollack; and in mathematics, Drs. John Chu and Michael Tabor. Sections of the manuscript have been read by some of the above, as well as by Drs. Jeremy Bernstein, Isaac Levi, Sidney Morgenbesser, Ernest Nagel, Barbara Sakitt, and Hao Wang. I am grateful to all of them for suggestions that helped to clarify my prose and eliminate errors. I would also like to thank Alice Mayhew, David Masello, and Catherine Shaw for their very helpful editorial suggestions, and James Danella for doing the line drawings. I thank Debbie Posner, the copy editor, for her help as well.

To Barbara
For her suggestions and inspiration

Contents

Introduction

In this book, I attempt to predict the changes that will take place over the next few decades in the content of science and in the lives of scientists. I concentrate on pure science, that is, the human effort to understand the universe in which we live and all it contains.

Science has progressed by first focusing on limited areas of experience and gradually extending the scope of the ideas developed to understand these areas. By this process, scientists have produced a coherent set of principles, with a wide range of applicability. Indeed, these scientific ideas now have no serious rivals for the systematic understanding of human experience. They form the core of the modern view of the world. New scientific principles soon influence other aspects of human thought. For example, the development of a materialist view of the world was largely a response to Newton's work on mechanics and gravity. We can thus expect that the new principles of future science will eventually affect all aspects of our world view.

There have already been immense changes in science over the past four decades. This period has seen substantial improvements in our understanding of many aspects of the universe, from subatomic particles through the molecular biology of heredity to the geological history of the earth. These changes in the content of science have developed with, and partly as the result of, great increases in the number of scientists, and in the level of public funding of science. So far as I know, no one predicted forty years ago that this rapid buildup in the scientific effort would occur, or what its ramifications would be. In retrospect, however, it is clear that changes in the content of science cannot be treated independently of the interaction of science with society.

Over the next forty years further important changes will take place in science. Should we, and can we, anticipate some of these changes? I would answer yes to both of these questions. The continued development of science is dependent on public funding of scientific research. The public is being asked to support future science rather than just to appreciate the results of past science, and the scale of support being requested is impressive. A new high-energy accelerator recently proposed by workers in particle physics would cost several billion dollars, and take over seven years to complete. Astronomers also have proposed expensive long-term projects, such as orbiting observatories. In order to justify such massive requests for funds, scientists need to determine where there are gaps in our knowledge and to anticipate the direction of future discoveries.

Efforts at anticipation can also serve a role within science. They can focus the attention of scientists on unsolved problems— and that can help lead to their earlier solution. Something like this happened in mathematics at the beginning of the twentieth century, when the German mathematician David Hilbert proposed a list of unsolved problems. Much of the work of mathematicians in this century has focused on the solutions to these problems, and others that grew out of them.

My analysis concentrates on two sciences, physics and biology. These two are the branches of contemporary science that are progressing the most rapidly, and many of the most important discoveries in the next few decades will be made in these sciences. By anticipating these discoveries, we will have a good idea of the shape of future science.

Physics deals with the most general properties of matter and energy. Its principles apply to all of nature, including those areas studied by the other sciences. Physics has made major advances in the past generation, especially in the understanding of the subatomic constituents of matter. These advances have led to further questions, which set much of the future agenda for physicists.

Physics is also the most autonomous science, in that its laws are unlikely to be derived from other sciences, as some laws of chemistry and geology can be derived from physics. Finally, the changes in professional lifestyle that have resulted from the increased expenditures and scope of scientific projects have been most pronounced in physics, so that by examining what is happening in physics, we can get some ideas of what soon may happen in other sciences.

In the future, physics will be able to provide detailed explanations where it can now only suggest broad outlines. This will come about largely through the introduction of new forms of mathematics and new computational tools that will allow for the analysis of various types of complex phenomena. This will allow for more concrete application of physics to other sciences dealing with such phenomena, including organic chemistry and biology. It will also extend the range of physical technology to such novelties as molecular engineering, and will allow for the construction of miniature devices of unparalleled complexity of function.

The subject matter of biology is as fascinating as physics, even though its laws are not applicable to as wide a range as phenomena. Furthermore, the great generalizations of biology, such as evolution, often have more immediate application to human thought in general than do those of physics. While biologists have succeeded in understanding many things over the last few decades, such as the biochemical basis of heredity, they have not yet answered many important questions, such as the origin of life or the cause of aging. There are good prospects for solving these problems over the next generation, through new theoretical and experimental innovations. Also, these developments could have immense practical consequences through their application in the new disciplines of biotechnology.

While physics and biology will come to resemble each other more over the next generation, leading to a more unified science, there is small reason to presume that biology will become a part of physics. Biologists generally do not doubt that the phenomena of life stem from the laws of physics. However, it is unlikely that we will soon be able to understand all of the activities of complex organisms through the application of these laws. It is even less likely that we will understand living things directly in terms of the subatomic particles of which they are made; too many of the details of

15

the phenomena are lost in the process when we try to find explanations across such a great gap.

The ideas of physical science have made some important inroads into biology (especially molecular biology) and there will be more such convergence in the future. Nonetheless, most of the phenomena of interest to biologists will continue to be studied through methods and ideas originating in biology, at least in the next few decades, so that biology, unlike chemistry, which is becoming a part of physics, will remain a relatively autonomous science.

Other sciences also study phenomena of great interest, but, like astronomy, they have effectively become a part of physics in their ideas and techniques, or like psychology, they have not yet progressed sufficiently for us to predict what their future will be. For these reasons, I do not concentrate on these other sciences in this book.

Predicting Scientific Change

To determine what science will be in the future, we need a conception of what science is. I define science as "the systematic effort to understand natural phenomena." Through this effort, scientists have spun a fine web of intellectual connections among these phenomena, and have discovered far-reaching generalizations—such as the law of conservation of energy—that play essential roles in many fields of science. It is these generalizations that most distinguish modern science from other attempts to find order underlying human experience.

Science is characterized as much by its institutions and its procedures as by its aims and results, all of which have evolved considerably since the sixteenth century, when modern science was born. In its modern form, the scientific effort has been highly cooperative, in that it combines the work of many scientists working on related problems. It has also been cumulative, in that the work of past generations of scientists serve as the basis for the present generation. These two features play fundamental roles in how science is and will be conducted.

There is no procedure for anticipating the future of science, no "science of science," and there are even scientists who argue that it is impossible in principle to do so. To show why I find their arguments unconvincing, I will describe several of the methods that I use to think about the future content of science. These methods are complementary in that they apply to different aspects of future science.

For those developments that will grow out of existing science, I try to identify the gaps in our present knowledge, then speculate on how they might be filled. It is somewhat like trying to surmise the shape of missing pieces in a jigsaw puzzle—an endeavor made easier when the puzzle is more advanced. This approach is a fruitful one for sciences such as particle physics or biochemistry, which have made great progress already. It is easier to predict the discovery of a new subatomic particle or a new chemical element than it is to guess the future achievements in a young field such as neuropsychology, where there are many missing pieces of knowledge.

In pursuing this "jigsaw puzzle" approach, I will describe (in Chapter 1 for physics and in Chapter 2 for biology) some of the crucial elements of our present state of understanding. These summaries are meant to make the book self-contained, so that a reader without detailed previous knowledge of some area of science will be able to follow the ensuing discussion of coming developments in that field. Because of its pragmatic function, I have restricted the summary to areas of science that I expect to change rapidly, such as molecular biology and cosmology, and omitted those such as acoustics and anatomy, where fewer new discoveries will be made.

I will focus, in Chapters 3 and 4, on areas in physics and biology where we can formulate questions to which we do not know the answers. In seeking answers to these questions much of the future development of science will occur.

Engines of Change in Science

It is especially difficult to predict developments that involve radical departures from the present state of science, since they are not hinted at by the current knowledge or theory. Yet these are the kinds of possibilities that particularly fascinate scientists and nonscientists alike. Radioactivity, for example, discovered in 1896, had been completely unexpected, and in order to understand it, it was necessary ultimately to go beyond Newtonian physics to quantum physics. In this book, I attempt to foretell such scientific developments by analyzing the ways in which science has evolved in the past, and using these insights to make educated guesses about future breakthroughs.

An important element in this analysis is to identify the main engines of change in science, the forces that act to change science from within and from the outside. One such engine is the development of new experimental techniques. This may take the form of novel methods for observation, such as the radio telescopes that revolutionized astronomy around 1950. These telescopes allowed astronomers to observe many new phenomena, such as pulsars, which led to radical changes in our picture of what the universe contains. Another form involves the application of experimental methods used in one science to the needs of other sciences. This is what happened when X-rays, which had been used by physicists to study the structure of metallic crystals, were adapted by biologists to the study of crystallized biological materials. Such studies led eventually to the unraveling of the structure of DNA.

A question raised by developments in experimental science is, To what extent does our picture of the universe depend on the instruments we have? Some scientists and philosophers, such as the British astrophysicist Arthur Eddington, have suggested that our conception of the universe relies critically on the tools we use for observing it, just as a fisherman's catch depends on the size of the mesh in his net. This view is too extreme. Scientists believe many things that they have not yet been able to observe for lack of instrumentation. For example, biologists believed in genes long before there were techniques to observe them directly. Currently, many physicists believe that the universe is filled with large numbers of subatomic particles called neutrinos, but these are not presently observable. The fact that such beliefs are often verified by later observations shows that we are not completely dependent on our present instruments.

Yet there remains an element of truth in the fisherman's net analogy, for there may be unimagined aspects of the universe that will be revealed only through the application of new tools for observation. Such discoveries have been made repeatedly in the history of science, especially in realms where there is little theoretical guidance as to what may exist—a situation that includes most of science. An early example was the discovery of microorganisms when microscopes were first focused on drops of water. A recent example was the discovery of integrated colonies of organisms living on the sea floor in regions where warm, chemically rich water wells up from the undersea crust. These form an ecology that is essentially independent of sunlight as an energy source, a possibility not considered previously for life on earth—even though there had been no theoretical reason to rule it out. A possible future discovery might be of microorganisms using chemical reactions in their metabolism very different from those we know. Such organisms might be discovered through a development as simple as a new method for growing microorganisms that will not grow in available media.

How scientists develop new methods of observation depends to some extent on the overall state of technology in the world. Sometimes a new technology is developed for its scientific applications, sometimes for its practical applications, and often for both reasons. For example, in the 1960s a series of discoveries in atomic physics culminated in the development of simple lasers. Using the results of discoveries in chemistry and optics, the laser's performance was improved in the hopes of finding practical applications. Eventually, the highly developed lasers were applied to new scientific problems, in atomic physics and particle physics.

In Chapter 5, I examine some of the new observational techniques that I expect to be developed in various sciences, and describe some of the phenomena they may uncover.

Scientific change is also driven by the application of novel forms of mathematics to science. Although twentieth-century mathematics has developed largely independently of natural science, there is some symbiosis between the two. It has often happened that some mathematical structures have proven to be the "natural" language required in a science, and important insights have emerged from the systematic use of these structures. This happened when Einstein adapted the mathematical structure known as tensor analysis for his general theory of relativity. Some biologists and mathematicians believe that a similar process is now taking place with the application of a branch of mathematics known as catastrophe theory to problems in biology. In the future, scientists will continue to make use of novel mathematics. In many areas of science there are phenomena so complex that they will require new symbolic descriptions if we are ever to understand them. I elaborate on the scientific applications of mathematics, and discuss the puzzling question of why mathematics should be applicable to science at all, in Chapter 6.

As scientific research has become increasingly dependent on public funds, society's attitude toward science has itself become an increasingly important engine of change. The manner in which

19

these funds are distributed—and the amounts available—are largely determined by factors outside the control of scientists. Yet these factors play a significant role in determining the scope of scientific research, and consequently the discoveries that scientists make.

It is unimaginable that the state of science would be anything like it is today were it not for the large investment of public funds into scientific research in the post–World War II period, first in the United States and then in other countries. The reasons for this investment were partly political, partly an expression of gratitude for the role of scientists in World War II, and partly obscure. There is no guarantee that the financial support will continue on a level that scientists consider desirable, and it is difficult to predict what will happen in this regard. Nevertheless, since well over half of the funds for basic research in the United States are provided by the federal government, some estimate of the future of public funding is essential to predicting the future of science. I record my own speculations about this in Chapter 9.

Society also influences science in ways not directly connected to the granting or withholding of funds. Most societies consider certain topics to be especially desirable for scientific investigation. Practical aims, such as the amelioration of the negative aspects of human life—the elimination of disease, for example—are always widely endorsed, and scientists pursue such goals unencumbered by public censure. Other scientific pursuits, especially those that challenge a society's general intellectual or moral opinion, can provoke great controversy. For example, Darwin's theory of evolution in the nineteenth century met with a vast public outcry, and continues to outrage some people today. When the implications of some scientific discovery are considered suspicious, society may attempt to limit research in the field. Nazi Germany, Stalinist Russia, and Italy under the medieval popes exemplify how social pressures can strangle scientific progress. But even in more benign societies scientific research can be impeded or "directed" by public opinion. For instance, in spite of strong support for any endeavor to cure individual diseases, many people (including some scientists) consider the search for a cure for aging to be objectionable, and this has probably limited progress in this field. Research on the causes of socially aberrant behavior, such as violent crime, has alternated between concentration on hereditary factors and on environmental factors, as social thought has varied.

The areas of inquiry that society regards as acceptable may have little to do with the state of science itself. Instead, these choices emerge from deep currents within the society. In order to predict the influence of these aspects of the future of science, one needs some sense of what may happen in the society as a whole, not an easy matter. Nevertheless, I have made some effort in Chapter 9 to guess what forms this influence may take.

In the last section of the book, I look at science as a social phenomenon. One aspect of this is applied science, or technology. In Chapter 8, I suggest some areas where developments in pure science, both in physics and in biology, may lead to novel forms of technology that would radically influence human life. I then discuss, in Chapter 9, how aspects of the professional lives of scientists, such as the way that they communicate their work to other scientists, will change as the result of developments within and without science.

The Processes of Change

A third method that I use to predict the future of science is a study of the internal processes through which science changes. Science proceeds by starting with a small area and progressing outward. Because of this process, there is nothing in our present scientific world that tells us how close we are to the boundaries of scientific knowledge. This is especially true in physics, which has no fixed subject matter, though it applies to the content of science as a whole. There are also branches of science that at first seem independent but that eventually merge with the rest of science. For example, the study of spectras—the light emitted by hot gases—became part of atomic physics. I expect the pattern of extension of the frontiers of science and elimination of internal boundaries between science to continue in the future. Some specific examples of this are discussed in Chapter 10.

Another relevant aspect of how science progresses is the relation between theory and experiment. Various views have been expressed on their relative roles, ranging from Newton's belief that scientists should follow precisely what experiments suggest, without making any extraneous hypotheses, to Einstein's dictum that the laws of physics are free creations of the mind, not directly obtainable from observation. I believe that the role of theory and experiment in science is more complex than these extremes suggest, and their relative importance differs from discipline to discipline and from time to time within each discipline. Theory plays a more important role in physics than in any other science, but the theoretical element in other sciences will become increasingly important.

The roles of theory and experiment in physics is best described as a game of intellectual leapfrog, in which first one and then the other play the leading role. There have been periods in which theories accounted for most of what was known; experiments simply tested new consequences of these theories or sought to discover unsuspected phenomena. The former was true for many years with regard to gravitation, where Einstein's general relativity theory prevailed. Progress has come from finding novel ways to test the theory, and by deriving unexpected consequences from it.

There have been other periods in which theories have been unsuccessful in explaining known phenomena. The main advances have come either from the creation of new theories, which were successful at explaining the phenomena, or from the experimental observation of new phenomena that provided the needed spur to further theoretical advance.

Experiment dominated in the period from 1950 to 1965 in subatomic physics; many new facts were discovered, but there was no adequate theory to explain them. Under the pressure of these experimental discoveries, the theorists eventually produced two important new ideas—the quark theory of hadrons and the unified theories of particle interactions. Theories based on these ideas explained most of the experimental discoveries of the previous two decades. But these new theories went further, and suggested that a large number of new phenomena had yet to be discovered because they would not be accessible through the available experimental methods. These implications were tantalizing, and challenged the experimental physicists to develop new techniques to observe them. They are meeting this challenge in the 1980s, and in the process will likely discover unexpected new phenomena that will put experiment in the lead again.

This model of the relation between theory and experiment suggests an approach to the prediction of future scientific discoveries. When a field of science has a theory that is generally successful in accounting for the known phenomena, one should look for the next discoveries among aspects of the theory that have not yet been verified. If the field is rich in phenomena that have not been successfully integrated into a general theory, one should expect new advances as a result of a theoretical synthesis. However, in order to carry out the synthesis, additional phenomena may need to be discovered first.

Theoretical analyses are often used in physics to give insight into situations which experiment may never be able to investigate directly. In recent years, through the methods of theoretical physics, scientists have studied processes which may occur in the very distant future, long after there are any intelligent beings left to observe them. Through such theoretical studies, the human mind can transcend some of the limitations placed on us by the fact that we are constructed of perishable matter.

All of this suggests that to predict the future of science, we must examine both theoretical and experimental innovations. I look at the former in Chapters 3 and 4 and the latter in Chapter 5.

The Importance of How We Calculate

In some branches of science, progress depends mainly on finding new ways to calculate, that is, to determine the consequences of what we already know in principle. This is presently the situation in the field of plasma physics, which is used both to study the conditions inside of stars and to attain controlled nuclear fusion on earth.

A plasma is a state of matter in which some of the electrons are permanently separated from atoms, and both the electrons and the charged atoms roam around freely. The behavior of plasmas is governed by well-known equations that have been studied in various contexts for many years. However, in the specific context of plasmas, especially when magnetic forces are also present, it is often difficult to extract from these equations accurate predictions about how the plasma will behave. Physicists firmly believe that this is a problem in mathematical computation, rather than of some flaw in the equations themselves. However, this distinction is small comfort to those trying to understand how a plasma will act under specific circumstances, or to predict the right conditions needed for useful nuclear fusion to occur.

In theoretical physics, the distinction between a fundamental innovation and mere progress in calculating the consequences of an existing theory is subtle. It is rare that a scientist begins a theoretical analysis with the notion that it will require revolutionary new ideas. Usually, scientists will begin with accepted knowledge and try to work out its consequences. When the result of a calculation differs from what is observed, it is difficult to decide whether to change the theory or the method of calculation, unless we know that the calculation offers an accurate picture of what the underlying theory implies. A case in point: Physicists believed for many years that their inability to make much progress toward a theory of subatomic particles was due to a lack of good calculational methods for dealing with their equations. After the equations were changed to conform with new ideas, however, it was found that the old calculational methods *were* adequate—provided they were applied to the right equations!

The influence of the method of calculation on the results that scientists obtain is so profound that particular methods often take on a life of their own; they may shape the overall picture that scientists have of some fields as much as or more than the basic ideas. In the late 1940s the American physicist Richard Feynman introduced a mathematical method for dealing with some of the equations of quantum field theory. His method involves a systematic use of certain pictures, now called Feynman diagrams, as a guide to how the calculations are done. Feynman showed how any interaction of subatomic particles could be represented by a series of pictures, and that, by applying simple rules to these pictures, the same mathematical results could be obtained as those that follow from the equations of quantum field theory. Even though he made no real change in the equations themselves, Feynman's methods proved so effective that his pictures have come to be a universal way of conceptualizing subatomic processes. Most physicists begin any consideration of these processes by drawing the relevant Feynman diagrams.

Calculations done with computers are beginning to radically influence how scientists pursue and envision their work. In many computer calculations, specific mathematical models of the phenomena being studied are introduced. In such models, some of the complexities of the phenomena are simplified, in order to allow for more convenient calculations. For example, objects that can actually move in three dimensions are allowed in some models to move only in one dimension. Sometimes, the specific assumptions made in the model come to be taken more seriously than was originally intended, effectively changing the fundamental laws so that we can calculate more effectively. While such simplification can lead to progress, we need to be skeptical of the implied notion that natural laws are so closely connected to the human ability to compute. If the main reason for introducing a model is its adaptability to numerical computation, it is unlikely that the model will survive the continuing rapid increase in computational ability. This is not to say that numerical computations should be regarded as automatically inferior to symbolic manipulation as a way of probing the implications of mathematical equations. Many important ideas have emerged from numerical calculations in recent years, both in science and in some fields of pure mathematics.

I expect this trend to increase as the use of computers becomes more widespread. For middle-aged scientists such as myself who have been trained to think of computers as relatively exotic innovations, it is a revelation to see how younger scientists, who have grown up with computers, have integrated them into their work. In Chapter 7 and again in Chapter 9 I discuss the role of computers in future science.

Part I
The State
of the Art:

What We Know Now

Matter
and Its Evolution:
The World of Physics

It is a common assumption that something qualifies as knowledge only if it is definitely and permanently known to be true, but this view is dangerously misleading. Science is an activity carried out by human beings, and infallibility is no more to be expected in what we do here than in any other human endeavor. Scientific knowledge should be understood as whatever scientists have good reason to believe at any given time. If one of these beliefs turns out to be wrong, then we can honor those who proved it wrong without disparaging those who thought it correct. That scientific knowledge is correctable is one of its great virtues. Indeed, one of the reasons modern science came about was in revolt against dogmatic religion, which made a virtue of constancy of belief. It would be a bitter irony if science were to repudiate this aspect of its origins by yearning for impossible certainties.

Some scientific ideas which are general in their applicability, such as the law of conservation of energy, will continue to endure. Such fundamental laws are seldom overthrown, though they are sometimes modified. As science progresses, the range of phenomena to which a fundamental law is applied may be restricted, or the meaning of some of the terms that enter into the law may be changed. This has happened in the past, and doubtless will happen again in the future. Those who use this evidence of scientific "imperfection" to criticize scientific knowledge as a whole are, to paraphrase Eddington, confusing science with omniscience.

There remain, however, important questions about what constitutes scientific knowledge. For example, in order to answer the question, What do scientists know now? one must first decide who is being referred to by the word "scientists." Is it a single eminent scientist, or a majority of those holding doctorates in a certain field, or members of the National Academy of Sciences? This is an especially troublesome question when one is considering topics on the frontiers of science, where different groups of scientists may have contradictory points of view. Even within those groups, it is likely there will be disputes over what constitutes the "truth." It is doubtful, for example, that any consensus could be expected on topics such as the cause of aging in mammals, or the origin of life.

There are areas of disagreement even within the more established areas of science. It is not at all clear, for example, which views about the foundations of statistical physics or quantum theory would receive general assent from the physicists in the National Academy of Sciences or from those on the faculties of the major world univerisities. An important reason for such persistent discord is that, in most cases, there are no formal procedures within science for reaching a general consensus about matters in dispute. Individual scientists make up their own minds about such matters, and decide what they will teach or investigate. These individual decisions rarely are converted into a formal position of science as a whole. Consequently, it is not easy to determine the present state of knowledge in any scientific field, especially those on the frontier.

Because of these complexities, no summation of the state of present scientific knowledge can be definitive. I have reached the conclusions put forth in this book by reading the publications of scientists currently working in various fields, and by discussing with prominent scientists their views of the present state and future prospects of their science. One flaw in this approach is that it omits the views of scientists who are outside the mainstream. This omission is not crucial for summarizing present knowledge, however. Even though some of the work of these maverick scientists will turn out to be correct and important, its general adoption is really part of future science.

This view of what scientists know is a subjective presentation. The reader should recognize that another scientist might have different views, and there is no "court of scientific appeal" to decide at present who is right. Only time, and the decisions of future scientists, will show whether what I now believe to be scientific knowledge qualifies for that distinction.

Physicists know the answers to two kinds of questions. One question is, What is the world made of?; that is, What are the relatively simple constituents that are the building blocks of matter? The second question is, What are the fundamental laws of nature?; that is, How do these simple constituents, and the more complex things made of them, behave under various conditions? Many physicists find it tempting to believe that in answering these two questions we can, in principle, provide explanations for everything of concern to science, not only in physics, but in biology and even psychology. Whatever the merits of this claim, it is not true in practice at present, nor is it an urgent objective within science. What is most important now is the extent to which our fundamental knowledge of constituents and laws can be used to explain the remaining phenomena of physics.

The World Is Made of Subatomic Particles

According to contemporary physicists, the world is made of several types of objects, collectively referred to as subatomic particles. (These particles can also be thought of as manifestations of something yet more fundamental, known as quantum fields. This is discussed later in this chapter.) There may be as many as 10^{89} identical copies of some of these particles in the present universe. The forms of matter familiar to us, both living and nonliving, on earth and in the heavens, are all composed of various combinations of only three types of subatomic particles—protons, neutrons, and electrons. Dozens of other types of particles can be produced momentarily in the laboratory, however, and are thought to have existed in large numbers in the early universe.

All subatomic particles are defined by a few qualities that they may possess, such as mass, spin, and electric charge. Two particles are of the same type if all of these qualities agree. Otherwise, they are considered to be different particles. Particles of the same type are, as far as we know, truly identical in these properties of mass, spin, and charge rather than just very similar. If all electrons were not identical, matter would collapse. If all photons, the particles that make up light, were not identical, lasers would not operate.

The subatomic particles readily convert into one another

when they collide. The kinetic energy of motion of light particles can be converted into the energy associated with mass (rest energy) of heavy particles. In many cases, even isolated particles can convert spontaneously into others, if the latter are less massive. In all such transformations, only a few properties, such as the total electric charge, remain unchanged. The subatomic particles do not act like the changeless building blocks imagined by some Greek philosophers. In the last few years, physicists have realized that even which subatomic particles exist has changed radically over the lifetime of the universe. It appears that evolution takes place on all levels of matter, not just on the more complex levels of living things. The driving force behind this evolution is the expansion of the universe, which by changing the environment in which particles are found, changes the particles themselves. Only twenty years ago, the idea that the properties of subatomic particles might depend on their environment would have been considered heresy. Nevertheless, there is now considerable theoretical support for this conclusion.

Under the conditions in which physicists usually observe subatomic particles, their defining properties are not perceived to vary, giving these properties an illusion of stability. However, under the immense temperatures and densities that prevailed in the early stages of the universe, the properties, such as mass, of some particles would have been very different from what they are now. The situation is akin to the variability of a liquid such as water. Under a fairly wide range of temperatures, water remains liquid, and its properties do not change much whatever the temperature within this range. But if the water is subjected to much lower temperatures, or is heated to above 100° Celsius, its properties change abruptly. The liquid becomes a solid (ice) or a gas (water vapor). This type of change, in which the properties of a substance change drastically as a result of a small variation in its environmental conditions, is called a phase change by physicists.

The presumed change in the properties of subatomic particles at very high temperatures is also considered to be a phase change, one that involves the properties of space, as well as of the particles in it. In other words, the particles do not react directly to a temperature change but to some alteration in space, the medium in which they find themselves. (This is similar to the various ways in which some chemicals react, depending on whether they are dissolved in liquid water or suspended in solid ice.)

It is easy to boil or freeze water, but very difficult to duplicate in the lab the extreme conditions present at the birth of the universe. Yet physicists have become convinced of the theory that subatomic particles, and space itself, went through momentous phase changes during and after the Big Bang. The rapid cooling that followed that primordial explosion is thought to have generated several phase changes. After an incredibly short time (perhaps a microsecond), the subatomic stuff of the young universe became stabilized, combining into the particles that make up matter today.

The Properties of Space

How is it possible for space to change, if space is conceived of as nothing at all? Actually, physicists no longer think of space in that way. Einstein, in his general theory of relativity, following up on the work of the nineteenth-century mathematicians Bernard Riemann and William Clifford, asserted that the properties of any region of space depend considerably on the presence and form of matter nearby. For example, the space near the sun is distorted in its geometrical properties; it "curves" because of the star's great mass. A triangle drawn by intersecting light rays near the sun would not obey the rules of Euclidean geometry; its angles would not add up to 180 degrees. It is this distortion of space and a related change in the way time passes that earlier physicists identified as the force of gravity, and which in Einstein's theory leads to the motion of the planets in orbits around the sun.

The idea that space can curve also applies to one of the great unanswered questions in astrophysics, Is our universe finite or infinite? If the universe contains enough matter, it would distort space so much that the universe would be finite—much as the surface of a sphere is finite (whereas a plane surface is considered infinite). This hypothetical distortion in the shape of the universe due to the presence of matter is similar to the distortion in the shape of a rubber sheet due to a weight that is lying on it: The amount of distortion increases as the weight of matter increases.

If the universe is finite, its overall size, compared to the size of some of the objects it contains, can vary over time, just as the size of a balloon can be inflated or deflated without changing its basic shape. There is convincing evidence, from a study of the light emitted to us by distant galaxies, that our universe is expanding. This does not prove that the universe is finite, however, and merely inflating like a toy balloon. Even if the universe were infinite, such an expansion could be understood as a change in the average distance between objects within the universe [Figure 1]. In either case, the expansion implies that the scale of the universe was much smaller in the past, and the objects within it were once much closer together. This discovery, the expansion of the universe, was first made by the American astronomer Edwin Hubble in the 1920s.

There is another way in which the properties of space can vary. Recent discoveries in subatomic physics have led physicists to think that space devoid of observable matter may, nevertheless, contain varying amounts of entities known as quantum fields.

Quantum fields are the modern versions of the classical fields introduced in the nineteenth century to account for such phenomena as electric and magnetic forces. The classical idea was that one electrically charged particle produced a field in its vicinity. This field, in turn, produced a field slightly farther away from the particle, which eventually reached the vicinity of a second particle and influenced its motion. This sequence of events produced the effect that we identify as forces. In this way, forces could be understood as having only local effects, and physicists could avoid the idea that a particle could act directly on another across large distances.

Such classical fields, like all physical properties in pre-quantum physics, had definite numerical values; either there was a field present in some region of space and time or there was not. If no forces acted in the region, the field had a numerical value of zero. If forces acted in the region, the field had some measurable nonzero value, and this numerical value changed continuously through space and time. Classical physicists believed that every event in the physical universe was potentially measurable and predictable.

In quantum physics, the situation is not so simple. Today's physicists hold that it is actually impossible to measure a precise number for all physical quantities at the same time—that the knowledge of certain quantities *precludes* the knowledge of others. This is the general content of what has come to be known as Heisenberg's uncertainty principle.

One consequence of this principle is that we cannot specify precisely both what fields and which particles are present in some region of space. Either the value of the field everywhere can be

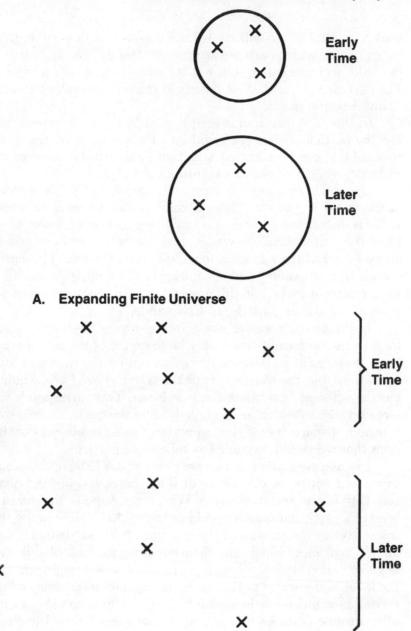

A. Expanding Finite Universe

B. Expanding Infinite Universe

FIGURE 1 Expanding Universe. Both a finite universe, represented by A, and an infinite universe, represented by B, can expand so that objects within each get ever farther apart.

known, in which case nothing is known of the particle content, or we can know which particles are present and the average value of the field over the region, but not its precise value at each point. The particle content and average field value is generally the more useful description.

In this new quantum description of fields, the classical field and the particles have been fused into a single concept, the quantum field. It can be said that quantum fields are the core reality, and that everything else is a manifestation of them.

One might compare the state of the quantum field to the level of the water in the sea. There is some reference level, in which water is distributed more or less uniformly. However, when winds blow the water develops waves, and the water level would be above or below the reference level at different places. This corresponds to a situation in which the quantum field is "excited" in some region of space, and this appears to us as the presence of one or more subatomic particles in that region.

In the analogy we are using, it is also possible for the water level to rise uniformly over a very large region of the sea, as when the tide comes in. In this case there may not be any waves, or local excitations, but the distribution of the water viewed as a whole is clearly different from when the tide is out. This corresponds to a circumstance in which the average level of the quantum field over a region of space is high, but in which there are no local excitations that we would recognize as subatomic particles.

Just as a swimmer is not aware of the water level in the ocean, so we will not be readily aware of the average level of the quantum field in any region of space. We cannot hope to measure the level of a particular quantum field unless we have some probe that is sensitive to the amount of the quantum field that is there. Subatomic particles, which are themselves excitations of quantum fields of various types, are such probes. Just as we might measure the level of the water in the sea by letting out an anchor and observing how much line is needed to reach bottom, so the presence of subatomic particles in a region of space can be used to probe the amounts of the various quantum fields in that region of space.

This can be done because the properties of the particles depend on the average level of the quantum field in which they are immersed. Several different subatomic particles might display similar properties in a region of space where a specific average level of quantum exists. In another region, where the level of field is different, the particle properties may become different from one another. For example, electrons and quarks have a very different mass in our region of space-time, but would have had equal masses in the conditions that prevailed in the early universe.

This is analogous to the behavior of colored objects when illuminated by different qualities of light. Imagine several objects that are similar except that they reflect light of different colors. If they are all illuminated with white light, an equal mixture of all colors, the objects will appear equally bright. However, if they are illuminated with light of a single color, red, only those objects that reflect red light will show up; the others will be invisible.

The circumstance in which particles that would otherwise have the same properties differ because they are immersed in some quantum field is one aspect of what physicists call "broken symmetry." Broken symmetry has turned out to be an important key to understanding the subatomic particles. Instead of having to study many types of particles, and to introduce a variety of distinct properties, such as different masses, into the equations that describe them, we can hope to see these differences arise in a simple theory, as the result of an asymmetric environment.

The broken symmetry of the properties of particles is a consequence of a broken symmetry of the underlying quantum field. The equations that describe quantum fields are thought to be symmetric; there are simple mathematical relations between the equations describing different fields, such as those associated with quarks and those associated with electrons. However, physicists have realized over the past twenty years that many of these equations have solutions that are not symmetric. These solutions correspond to average levels of the quantum field in some region of space that is different for one field than for another. When this situation applies in some region, the symmetry for those fields is said to be broken. Because these average field values influence the properties of any particles present in the region, these particles may also be observed to differ, even though they are described by similar equations.

One advantage of using symmetric equations for the quantum fields is that there are usually many fewer symmetric theories than unsymmetric ones, and the symmetric theories often have properties that make them more tractable mathematically.

Particles and Fields

Unfortunately, it is beyond our capability to test the idea of broken symmetry by changing the value of quantum fields over a large region of space at will; too much energy is required. It is conceivable that someday we will be able to accomplish this over small regions of space, through collisions of heavy nuclei at high energy. However, as stated earlier, scientists recognize that such changes in the state of empty space took place as a matter of course in the very early history of the universe.

If we could study this history through observation of its effects on features of the present universe, we could learn about properties of subatomic particles under conditions very different from any that we can produce in our laboratories. Such a study is just beginning, and will be an important part of future physics. Some aspects of it are discussed in Chapter 5. Meanwhile, theoretical physicists have worked out, in some detail, the history of phase changes in the early universe.

These early phase changes were a consequence of the rapid expansion of the universe. During this expansion, the temperature of the universe continuously decreased. This is similar to what happens when a gas of atoms expands and cools. Since the state of quantum fields depends on temperature, as the universe expanded the quantum fields that occupied it underwent various phase changes. In these phase changes, the average value of the field changed abruptly, leading to a corresponding change in the properties of any particles present as well. For example, the mass of a particle representing excitations of a specific quantum field might have changed from zero to some large value. As the phase changes take place, there is a complicated exchange between the energy of the field, the energy of particles, and the expansion energy of the universe. This is similar to what happens during the liquefication phase change of an ordinary substance, where the energy difference between the gas and the liquid form of the substance is released as heat. The energy released in phase changes of quantum fields can trigger the creation of particles, so that the number of particles present in the early universe also changed abruptly.

Because of these early changes in the properties and numbers of particles, one could, theoretically, make several different lists of the types of subatomic particles that existed in various periods in the history of the universe. (This problem is similar to that of making a list of the animals present on earth at different times; the list would vary considerably over the course of evolution.) The simplest list would be of those particles that existed in the earliest stages of the universe, when its overall temperature was some 10^{28}

times higher than it is today (three degrees above absolute zero). At that high temperature, space was in a phase in which each quantum field, except perhaps that associated with gravity, had a minimum value everywhere in space. In this situation, there were no broken symmetries. All subatomic particles had zero mass and consequently traveled at the speed of light.

The number of distinct particle types that existed in the early universe, when symmetries were unbroken, was quite large: there were at least fifty-seven and perhaps substantially more. This estimate is based on the similar particles we know to exist in the present universe—yet we cannot be sure of the original number, for a reason that will appear shortly. The different particles were then distinguished from one another only by electric charge—or by other properties similar to electric charge—and by spin. In this stage of the early universe the particles of each type were present in essentially equal numbers because each type was readily created in collisions between other types, so any imbalance would quickly be removed. This differs from the present situation, where the number of particles of each type is the result of a complicated history.

As the universe continued to expand and cool, the quantum fields went through several distinct phase transitions. Each of these led to a change in the overall level of at least one quantum field everywhere in space, and an associated change in the properties of some subatomic particles. The last of the phase changes is thought to have taken place when the temperature of the universe was about 10^{13} times as great as today. After that, all subatomic particles had the same properties that we find them to have now.

According to theoretical analyses, all of these extraordinary changes took place within a very short period of time—perhaps in the first microsecond or so after the expansion of the universe began. In other words, the main subatomic features of our universe were determined in a flash of time, and the consequences have been working themselves out ever since. As Omar Khayyam wrote (in William Fitzgerald's translation), ". . . the first morning of creation wrote what the last dawn of reckoning shall read."

Let us then jump in time to the present universe, some 15 billion years after the microsecond we have just described. The same types of particles potentially still exist, in the sense that the same quantum fields can still be excited. But now these particles differ drastically from one another in their masses, most of which are no longer zero. This change has led to radical differences in the number of particles of any given type that are actually present.

One reason for this is that in our current "phase," the particles with greater mass are usually unstable against decay; that is, they tend to transform spontaneously into particles of smaller mass. Even some of the particles not subject to decay because of their low mass, such as electrons, occur in much smaller relative numbers than they once did. This is because most of the electrons were annihilated long ago together with their anti-particles, the positrons, leaving only a small remnant of survivors. The only particles thought to exist in numbers comparable to those of early times are those particles with zero or very small mass. Particles of this type include photons, the particles that make up light, and three types of electrically neutral particles called neutrinos.

Most of the particle types that were abundant in the early universe have long ago disappeared. We only observe them when they are produced briefly in laboratories, and then annihilate or decay. Because of this, we are uncertain of how many particle types may exist. Particles with large mass and short lifetime would not have survived, and could only be found if produced recently. We know of several such particles, including muons, which are particles similar to electrons but two hundred times more massive. But there may well be other types, yet unknown to us, which are so massive that they cannot be produced by any existing accelerators. Such presently unknown particle types might have existed in large numbers in the early universe. For one thing, the average energy of the particles present then was so high that even very massive particles could have been produced in collisions. Also, particles that would have very high mass if produced today could have had small or zero mass in the early universe, because some phase transitions that would give them mass had not yet occurred. Some theories of subatomic particles imply that many such presently unknown particles did exist in the early universe. It is possible that some of them have properties that enabled them to survive the decay and annihilation of most of their brethren, and have survived as fossils into the present universe; and we may find them by suitable searches.

The World of Quarks and Electrons

Some of the original particle types, quarks and electrons, still occur in relatively small numbers in the present universe. There were slightly more particles than antiparticles present originally, and this small excess survived the annihilation. At least five quark types existed in the early universe, but only two of these—u-quarks and d-quarks—have survived. Although all of the matter that is in our bodies, in the earth, and in the stars is made of quarks and electrons, in terms of sheer numbers, quarks and electrons occur only about one one-billionth as often as do photons and neutrinos. The matter that most interests us is no more than a small contaminant in the overall matter balance of the universe.

In the present universe, quarks and electrons have properties that allow them to form the tightly bound clusters that we call nuclei and atoms. Photons and neutrinos cannot do this, and so exist much more diffusely throughout the universe. The situation is a bit like comparing bacterial cells with elephant cells. There are many more of the former on earth than of the latter. But the elephant cells are only found in the high concentrations that we call elephants, whereas the bacteria occur individually, so we are much less aware of them.

Nevertheless, most of the universe we know is made of quarks and electrons, and the present picture we have of the world is largely an expression of the properties of these particles. Of the two, quarks have a greater tendency to cluster together. Indeed, this tendency is so pronounced that most physicists believe that quarks are never found in isolation, but only in combinations containing either three quarks or one quark and one antiquark (the antiparticle of the quarks). These are the combinations that make up most of the subatomic particles that we observe, such as protons and neutrons, the particles found in the nuclei of atoms.

The reasons why quarks insist on clustering in this way are not completely understood. There is a general theory, known as quantum chromodynamics (QCD), that attempts to describe how quarks behave. QCD involves the interactions of fields associated with quarks and fields associated with another type of particle called gluons (so named because they bind the quarks together). Most physicists believe that when the predictions of this theory are better understood, we will know why quarks cluster as they do. This is an example of a type of problem mentioned in the Introduction: that of working out the precise implications of a generally accepted theory. It is possible, for instance, that the tendency for quarks to cluster is not universal; at least one group of physicists has reported observing what could be isolated quarks, but this report has not been confirmed.

Ever since the first microsecond after the origin of the universe, quarks have been bound together, in groups of three, into neutrons and protons. All of the other combinations of quarks, either of u and d quarks, or of the other quark types, which also can bind together, are unstable under present conditions. That is, if they are produced, they change spontaneously into less massive particles, and eventually into some combination of the stable ones. Even neutrons are unstable when they are found in isolation—as when they are produced in nuclear reactors—and decay into protons in a few minutes. The reason that neutrons exist at all in the present universe is that when given the chance, they bind together into more complex and lasting objects. Neutrons can bind with protons into the objects that we know as atomic nuclei, and with one another in immense numbers into neutron stars.

Electrons also bind with nuclei and with each other into the combinations that we know as atoms and molecules. This binding occurs through electric and magnetic forces, which are manifestations of the same quantum field whose particle aspect is the photon. The detailed properties of this field are summarized in a theory known as quantum electrodynamics (QED), the most widely tested theory in quantum physics. No inaccuracies have been found in the theory, down to a level of error of less than one part in a billion.

Most physicists believe, on the basis of theoretical arguments, that even protons and bound neutrons are not really stable, and that over sufficiently long periods of time they decay into electrons or neutrinos. Such decays have not yet been observed, although experimental searches are under way. The time period over which this is thought to occur is 10^{31} years or more, so that few of the protons and neutrons produced in the early universe would have decayed yet in this way. However, by looking at matter containing thousands of tons of protons, a few proton decays should be seen in a year. According to this theory, if the universe continues to expand for another 10^{31} years or more, matter as we know it will have disappeared. The era in which the universe is dominated by the matter familiar to us will be very long by human and by galactic standards, but it may still be just an instant in the whole history of the universe.

How Things Happen in the Quantum Universe

There are a few basic principles that govern the behavior of quantum fields, the particles that are associated with them, and everything that is made up of these particles. These principles fit into a general theory called relativistic quantum mechanics, which was invented in a short period between about 1925 and 1935. This theory was applied first to particles and then to fields. Ever since then, physicists have been living off the intellectual capital of that rich decade, applying the principles of relativistic quantum mechanics to various entities, and sharpening their methods for extracting the consequences of the basic ideas. No fundamentally new laws have been found necessary to describe any of the phenomena that physicists have discovered in the intervening years.

Relativistic quantum mechanics involves several types of principles. Fundamentally, the theory describes the *probability* that any type of measurement performed on a physical system will give a specified result. Only in special cases does this theory make definite predictions about physical objects.

The principles of relativistic quantum mechanics relate measurements done under one condition to measurements done under another. For instance, one principle relates the results of measurements done on a physical system at one time to those of measurements done at another time. If an electron in an atom is measured to have a certain velocity at one time, then relativistic quantum mechanics predicts specific probabilities that the electron will have one or another different velocity a second later.

A startling feature of quantum theory is the implication that measurements can interfere with one another; that in order to carry out one type of measurement it may be necessary to set up experimental conditions that destroy the information gained as the result of a previous measurement. In the example just given, the precise measurement of the electron's velocity would necessarily destroy any previous information about the electron's location. This is because any method by which the velocity of an object can be measured necessarily changes the location of the object in an unpredictable way. For example, if we measure the velocity of an electron by a kind of radar, by shining light on the electron and observing the change in wavelength of the reflected light, the effect of the light will change the location of the electron in an unpredictable way. This type of interference between measurements is a key element in Heisenberg's uncertainty principle, which limits the accuracy with which we can know the present state of any physical system—and hence also limits the accuracy with which we can predict its future behavior [Figure 2]. For everyday objects, the limitations due to Heisenberg's principle are too

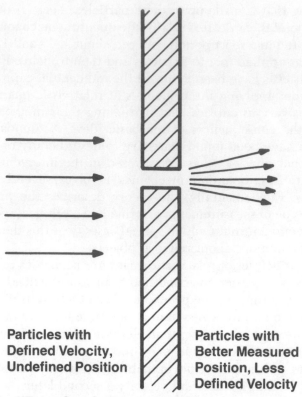

Particles with Defined Velocity, Undefined Position

Particles with Better Measured Position, Less Defined Velocity

FIGURE 2   Heisenberg's Uncertainty Principle. A measuring device, such as the small hole in the screen, which gives us more knowledge of the position of particles in a beam, causes a loss of knowledge of the velocities of these same particles, because of its interaction with them.

small to be important, but for atoms and subatomic particles they are crucial.

Paradoxically, this very uncertainty has led to equations that in some circumstances are far more precise than those used by the classical physicists, who were so confident of the predictability of natural law. These quantum equations sometimes imply that a physical quantity such as energy will always be found to have one of a small number of numerical values, rather than an arbitrary value. One way in which the mathematical equations of quantum theory can be obtained is by taking an existing set of equations from a pre-quantum theory and modifying it to include the mutual interference of measurements. The modified equations describe the same phenomenon that the classical equations described; the difference is that the quantum mechanical equations are correct.

Understanding subatomic particles as excitations of quantum fields is a specific consequence of the application of this process to the equations that describe classical fields. The quantum equations have solutions that describe the unique properties of the subatomic principles, such as definite electric charge and spin, whereas the equations for the classical fields did not. (More recently, it has been found that these same equations for quantum fields also have other solutions, which may describe novel objects distinct from subatomic particles.)

It is possible to reverse the procedure just described. One can begin with the properties of the subatomic particles and introduce quantum fields as a convenient mathematical description for these particles. For many years, this alternative notion—that particles were the primary reality and that fields were just a mathematical convenience—was the one that most physicists, including myself, accepted. More recently, however, the concept of broken symmetry, which relies on the notion that quantum fields pervade space, has changed this attitude, and has led many physicists to think of the fields as primary, and the particles as secondary manifestations. This swinging of the pendulum between field and particle is one aspect of a wider swing between continuity and discreteness—which has been going on in physical theory for two millennia.

The invention of relativistic quantum mechanics required a concerted effort to incorporate into it the requirements of Einstein's special relativity theory, which established the relativity of space and time. When these requirements are imposed, they lead to mathematical formulas that relate the measurements made on the same physical system by different observers, such as one on earth and another in a spaceship moving at constant speed. Such mathematical relations strongly limit the form that any theory can have, because most imaginable theories would not be consistent with them. One spectacular example of this was the conclusion, first recognized about fifty years ago, that any theory describing quantum fields that satisfies the requirements of special relativity necessarily allows for the creation and destruction of particles of one type out of other types. This prediction came just in time to account for the wide variety of particle creation and destruction phenomena observed in laboratories in the past fifty years.

Many physicists believe that a gross deviation from the description of space-time used in the existing form of relativistic quantum mechanics would come from extending that theory to include the requirements of Einstein's other great creation, general relativity, which describes the influence of acceleration on observations. Most physicists now think of general relativity as mainly a theory of gravity, one of the ways in which particles, or things made of them, can influence one another. Gravity also plays a special role in nature, because its effect is in one sense universal, affecting everything in a similar way. One manifestation of this universality was Galileo's discovery that all bodies fall to earth with approximately the same acceleration. The influence of gravity affects even the instruments with which we measure distance and time, leading to the curvature of space-time recognized by Einstein. For example, similar clocks, one placed on earth and the other near the surface of a neutron star, would run at very different rates.

But what about the effects of gravity in very small regions, whose size is 10^{-33} centimeters or less? We are now considering distances that bear the same ratio (10^{-20}) to the size of a proton, itself usually thought of as tiny, as the size of a proton does to the size of a large mountain. At these minute distances, the effects of quantum theory on gravity become important—yet physicists have not been able to formulate a theory to describe what happens. They do believe that such a theory would describe properties of space-time radically different from those that operate over the much larger regions studied heretofore. This is one of the areas in which scientists expect to make significant progress in the future, and I return to it in the next chapter.

In its present form, relativistic quantum mechanics is not a complete physical theory. Among other things, a complete theory would have to determine what entities actually exist in the universe as well as to specify the behavior of those entities that do exist, and the existing theory does not do this. What physicists do at present is to supplement the basic principles of the theory with an empirically derived list of entities to which it should be applied. One way to do this is to list all of the known subatomic particles, together with those believed to exist but not yet discovered. The latter could include particles that are now too massive to be produced in laboratory experiments, but which existed in the early universe. A more effective way of describing the objects to which relativistic quantum mechanics is applied is by a list of the types of quantum fields, since the properties of the fields change less during the evolution of the universe than do those of the particles, and because physicists now think of the fields as the more fundamental entities.

The list includes those fields that account for known particles, as well as other fields whose existence we must assume because of an important aspect of our description of particles and fields known as internal symmetry. An internal symmetry is a mathematical statement relating the properties of different entities at the same space-time point. For example, one such symmetry might state that the quantum field used to describe electrons satisfies the same equations as the field used to describe one species of quark. It is hypothetical relationships of this type that have led to theories implying that protons sometimes decay into positrons. Internal symmetries are distinguished from space-time symmetries, such as those implied by relativity theory, which relate fields at different locations in space-time.

In some cases, all of the objects whose properties are related by the mathematical statement of a specific internal symmetry are already known to exist, and it is not necessary to assume any additional entities. The symmetry relating one quark field to the electron field is an example of this. In other cases, in order for an internal symmetry to apply in nature, quantum fields not yet discovered must exist in addition to those already known. The logic of this argument is similar to that used in deducing the existence of volume 23 of an encyclopedia from the observation of volumes 1–22 and 24–30. In many cases, physicists are convinced of the applicability of an internal symmetry by studying known particles and comparing their properties to those predicted by the symmetry. Based on such confidence about the truth of the symmetry in nature, this argument can then be used to predict the existence of previously unknown fields and their associated particles.

Such predictions have been made several times in the past few decades. For example, in the early 1960s, the American theoretical physicist Murray Gell-Mann inferred, on the basis of an internal symmetry that seemed to apply to several of the particles then known, that a then unknown particle with certain definite properties of spin, charge, and mass had to exist. His prediction led to the discovery of this particle, omega-minus, in 1964. The use of internal symmetries remains one of the best tools in the kit used by theoretical physicists to anticipate novel discoveries.

Ideally, we might hope to hit upon a principle of internal symmetry so all-encompassing that it would determine every type of quantum field in existence. Physicists working on this problem make liberal use of a branch of mathematics known as group theory. A group is a collection of mathematical objects—the members of the group—and a set of rules that describe how any two of the members can be combined to produce a new member. It is these rules of combination that distinguish one group from another. The same rules can be applied to very different sets of objects, such as numbers, geometrical figures, or quantum fields. Each such application gives a different illustration of what mathematicians consider to be the same group, somewhat in the way that different families can illustrate the same rules for determining kinship relations. Mathematicians have studied the properties of groups in great detail, and much information about them is available to physicists. (Groups are also used in physics in other ways than to describe internal symmetries—for example, to make precise the relations between observations made by two people in relative motion.)

As physicists use groups in quantum field theory, each member of the group can be thought of as a specific mathematical relation among the properties of a number of quantized fields. For example, one member of a group might be the mathematical operation that changes an electron field into a neutrino field, and vice versa. If physicists knew which group applied to the internal symmetry of quantum fields, they could say a lot about which quantum fields exist. This is because for any specific group, the mathematical relations that are its members must apply to a definite number of fields. We can say that each group corresponds to a different number of slots that can be filled by quantum fields. And if we knew which fields exist, we could then say something about which particles exist.

For example, quantum chromodynamics—the theory relating the interaction of quark fields and gluon fields—is based on a specific group thought to describe some of the properties of quarks. The rules of combination for this group require quarks to exist in multiples of three; the rules could not be satisfied by any other constellation of quarks. This type of restriction of physical possibilities as a result of the application of specific mathematics is not new in physics. It is analogous to Isaac Newton's demonstration that the application of his law of gravity to the solar system implied that the orbit of any object could only be one of three possibilities: an ellipse, a parabola, or a hyperbola.

The group that is used to describe quarks was introduced after it was known that quarks come in groups of three, so the existence of quarks in groups of three was not a prediction. In other cases, physicists first postulated that some specific group was the appropriate one, and then examined it to see what combination of fields and particles it allows to exist. Remarkably, proceeding in this somewhat haphazard way physicists have been able to tease out some of the internal symmetries of nature. However, until we have some general principle to determine what internal symmetry groups are applicable in physics, there will be an unavoidable empirical element in this approach to determining what actually exists in nature. The theory will not be complete in the sense described above. This is one of the most active fields of research in particle physics today.

What Can Be Understood From Particle Physics

Physics consists of a set of principles that are widely applicable to a collection of fundamental objects. An essential question that any science must face is the extent to which the phenomena within its scope can be adequately understood by making inferences from the fundamental principles of that science. For physics, this question takes the following form: To what extent can the known properties and behavior of matter be understood by applying the laws of relativistic quantum mechanics to quantum fields and their particle manifestations?

The answer is, to a large extent! Explanations in principle exist for most of the known properties of matter, both in the conditions found on earth and in the conditions prevalent in other parts of the universe. The exception is, perhaps, for phenomena thought to have occurred in the earliest instants of the universe, and for those inaccessible phenomena that may occur inside of black holes. In some instances, in order to describe aspects of the behavior of bulk matter, it is necessary to make additional assumptions that cannot yet be deduced from the basic principles, although there is reason to think that this deduction should eventually be possible. For example, it has not been shown conclusively that the form of matter that we recognize as a solid body is the most stable arrangement of a collection of atoms at low temperatures. It should be possible to demonstrate this by applying the principles of relativistic quantum mechanics to the electrons and nuclei that make up the atoms, but there are severe mathematical problems in doing so. However, physicists have made a great deal of progress in understanding the behavior of solid bodies by assuming that they are indeed stable, and then studying how quantum theory applies to them.

For example, the way that metals and other solids conduct heat and electricity can be well understood by applying quantum theory to the behavior of electrons inside the solid. The eerie behavior of liquid helium at very low temperatures—it flows without any resistance, even over the edge of a container—can be understood by applying quantum theory to the atoms in the helium. The properties of the atomic nucleus, once thought to be beyond the reach of quantum theory, have yielded to a combination of simple models of the way the neutrons and protons in the nucleus influence one another and the application relativistic quantum mechanics to these models. The way in which light is emitted and absorbed by various bodies has been understood through the application of relativistic quantum mechanics both to the photons that make up the light and to the atoms that make up the bodies. In each of these applications, it was necessary to introduce certain mathematical approximations in order to make the phenomena amenable to our limited mathematical abilities.

This illustrates a general problem that arises in trying to do physics "from the ground up." Even when the basic principles in some field are very well understood, there are usually problems in applying them to complex phenomena. While relativistic quantum mechanics can readily be used to understand the properties of isolated simple atoms and molecules, it becomes prohibitively difficult to apply this theory directly to moderately complex objects, such as the macromolecules of interest to biologists, or even to groups of a few atoms. The problem lies not in the objects, but in ourselves: We are just not able at this time to draw the necessary inferences from the basic principles. What we usually do instead is to make simple models of the complex systems—models we believe to be consistent with basic principles—and then reason with the aid of these models about the phenomena that we wish to understand.

One example of this methodology is the biochemists' model of complex molecules, in which the atoms comprising the molecules are represented as featureless balls with protruding "hooks" that enable the balls to form chemical bonds. This model has been quite successful in representing much of the known behavior of organic molecules, but it cannot be taken as anything that resembles the way in which relativistic quantum mechanics would describe the same molecules. There is nothing wrong with using such models, but in order to make science into the seamless web of explanation we would like it to be, it will be necessary to fill in the gaps between the modes of explanation used by scientists working at different levels of complexity, such as particle physicists and organic chemists.

The application of relativistic quantum mechanics to the understanding of phenomena on a scale larger than the subatomic has been successful up to a point. When it has been unsuccessful, the reason probably has less to do with unknown laws than with the complexity of the problem; sometimes there are just too many things going on for us to unravel by methods presently available to us. The development of new models to do so is one of the important tasks for future science.

getting outside —
anti-organic
move at end of figures to
see from outside normal
viewpoint

2

An Awesome Multiplicity: The Science of Life

for 19cent —
change from organicism
to other model - change involved of method -
pattern of science
modelling on one kind of another
instead of another?

Gauguin -
spirit genesis - to general/personal
from empirical pattern to
symbolic pattern

Biologists work to understand living things—their structure, behavior, and makeup. More closely than physicists, biologists have followed the injunction of Francis Bacon about how science should proceed. Biologists begin typically with the observed phenomena of life, and then proceed from these observations to devise theories that encompass as many of the phenomena as possible. They seldom theorize about unknown possibilities.

For example, there has been little effort among biologists to anticipate the existence of living things apart from those that have been observed. Of course, generalizations do exist in biology, such as the ideas of evolution and natural selection, which have transformed our way of thinking in biology and in other sciences. However, even such powerful ideas as evolution have usually been used to account for known phenomena rather than to predict new ones. That is why this summary of what biologists know now is chiefly a description of known phenomena and the ways they are studied and understood.

Similar Structure Underlies Diversity Among Living Things

What is perhaps most striking about living things on earth is their diversity. Millions of species exist, each with distinct charac-

from biology to physics
19 cent → 20 cent

teristics. Among the multicellular creatures, even the different members of a species are recognizable individuals. It is not surprising that this appreciation of diversity should be our untutored response to living things; for most of human history, our lives have depended literally on being able to distinguish what we could eat from what could eat us.

As a result of long study, biologists have become convinced that underlying the diversity of life there are fundamental similarities among all the living things we know. A small number of chemical substances, and a larger but still restricted number of reactions among these substances, are the basis of all of the structure and behavior of living things. The diversity of life is analogous to the many different musical compositions that can be formed from a small number of notes and chords. It does not arise from differences in basic structure among the ultimate individual components of living things, but rather is an indication of the immense number of distinct patterns that can result from combinations of a large number of copies of a few distinct objects.

This process occurs on several different levels of structure in living things. On the most primitive level, all of the molecules found in living things are combinations of a small number of atoms, primarily carbon, hydrogen, nitrogen, oxygen, phosphorus, and sulfur. Out of these atoms, untold trillions of different small molecules can be formed. Yet life on earth makes use of a minuscule fraction of these possible molecules, probably no more than a few thousand, to carry out all of its chemistry. That all the activities of life can arise from combinations of this small set of molecules should be no more surprising than that all of the glories of color can emerge from combinations of three primary hues.

An even smaller number of these biomolecules, about twenty-five, form the basis of two important types of molecular polymers— proteins and nucleic acids. Twenty different amino acids, in various combinations, are strung together to form proteins, which act as catalysts, making it possible for most biochemical reactions to take place at rates that are great enough for living things to function.

In their biologically active state, proteins take complex three-dimensional forms. It is these precise forms that allow proteins to function as catalysts. It is not yet well understood how the three-dimensional structure of proteins is determined by their amino acid sequences. Proteins also perform other functions in living things, including forming parts of the structures that give each living thing its characteristic shape, and separating it from the environment. The precise way in which the amino acid sequence allows a specific protein to perform these functions is also unclear.

The other fundamental molecular polymers are the nucleic acids, DNA and RNA. In DNA, the units are a set of four bases, relatively simple organic molecules known as adenine (A), guanine (G), cytosine (C) and thymine (T). In RNA, a fifth base, uracil (U), substitutes for thymine. The bases are strung onto a long backbone, formed of sugar and phosphate molecules, somewhat like beads on a string [Figure 3]. DNA and RNA also take on complex three-dimensional shapes in their biologically active forms, and these shapes play important roles in biological activities. In most cellular living things, one form of nucleic acid, DNA, is used for information storage; the other form, RNA, for information transfer and the production of proteins. The rules for translating a sequence of bases into a sequence of amino acids appear to be nearly universal, and their elucidation was one of the major advances in the history of biology. The fact that there are two steps involved, rather than one, in going from genetic information to its protein realization, allows for elaborate control processes that determine how and when this information is expressed. These control mechanisms play important roles in the life processes of complex organisms, and biologists are just beginning to understand them.

It has been estimated that the DNA in the nucleus of a member cell of a complex organism, such as a human being, contains the information for the construction of many thousands of distinct proteins. Many of the proteins that are produced are the same from organism to organism. Different examples of an organism, or even different organisms within a general class, such as humans and chimpanzees, usually have very similar DNA sequences, in which only a few percent of the bases differ. A small variation in the proteins is enough to produce the great differences between two people, or between humans and apes. It is because of the exquisitely complex array of interactions between the molecules in a cell, and between the cells in an organism, that small variations in the units—molecules in one case, cells in the other—can become magnified into large differences. The mutual interaction of a few types of units is the underlying mechanism for the diversity of living things.

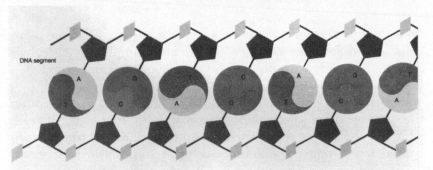

DNA segment

FIGURE 3   A Model of DNA. The phosphate and sugar backbone of each strand lie on the outside, while the bases, A,C,G, and T, lie inside, linked from one strand to the other according to the rule, A-T, G-C.

The similarity of living things extends beyond their chemical structure. All known life is made up of one or more cellular units. These cells consist of proteins and nucleic acids in close association, surrounded by some type of membranelike enclosure that keeps the cell contents together and keeps most of the environment from impinging. The cell membrane is not completely impermeable; if it were, the cell would have no way of ingesting needed energy from the outside or excreting its waste products.

The arrangement of the substances inside the cell is not uniform among all forms of life. In procaryotes—a wide class of simple living things that includes bacteria—the DNA is not isolated from the other contents of the cell by an enclosure. In eucaryotes—another wide class, one that includes all multicellular creatures—most of the DNA in each cell is found in a structure of its own, the nucleus, which is partially segregated from the rest of the cell by its own membrane. Protein and RNA are also found in the nucleus, where they exist with the DNA in complex arrangements that are crucial to the way eucaryotic organisms function. Other structures, called ribosomes, are also found in the cells of all living things. These are the "factories" in which proteins are synthesized according to the programs contained in the nucleic acids.

The simplest of cells are able to carry out many of the same basic chemical activities as the most complex. There are a few things that each living cell must do. These include: producing the proteins it needs from amino acids, transferring information from DNA to proteins, and duplicating the DNA itself when the cell divides. The structures within the cell and the sequence of operations used to carry out these tasks, are also almost the same in all cells. The differences between simple and complex living things have to do mainly with the more elaborate instructions encoded in the DNA of complex organisms, which enables them to produce a wider variety of proteins than the simple organisms. Furthermore, the complex organisms usually have many different types of cells, which, in addition to the general tasks that all cells carry out, are specialized to perform different chemical and physical functions.

Every form of life that we know embodies a different elaboration of a single basic idea: a symbiosis between nucleic acids and proteins. The elaborations are themselves of great interest, for human beings are one such elaboration. Just as physicists are not always able to understand the workings of complex systems composed of subatomic particles even though they know the laws that the individual particles obey, so biologists have had only partial success in translating what they know about the behavior and structure of single cells into a complete understanding of what complex organisms do.

How Diversity Has Come About

Millions of different species have existed over the history of life on earth, and within each species there have been very large numbers of individuals, related in some gross features but distinct in many ways. Even this diversity comes nowhere near to exhausting the viable possibilities that can arise from combinations of the basic biochemicals, any more than the millions of books in the Library of Congress come close to exhausting the possible range of meaningful English sentences.

There are two distinct problems involved in understanding biological diversity. One is the problem of mechanics, that is, how the differences in form can be understood in terms of the properties and behavior of the underlying nucleic acid and protein molecules. The other is the question of origins, that is, how this diversity came about. Life on earth presumably began with simple living things, and evolved over a period of several billion years into increasingly complex organisms.

It was Charles Darwin and Alfred Wallace who essentially answered the question of how the diversity of life on earth emerged. The present diversity has come about through the magnification of many small differences, called variations, over immense periods of time. Organisms that are very similar can have offspring that are different in some ways, and over many generations, these differences can increase so that the descendants are no longer part of one species. This can most easily happen when there are several groups of organisms reproducing in isolation from each other. The time scale over which such large differences appear is not precisely known and is the subject of some controversy among biologists, but there is little doubt that the general picture is correct.

Darwin did not know what caused these variations from organism to organism within a species. With our present model of how genetic information is stored, we know of many ways in which variations can arise, both as the result of environmental influences on the genetic material of the organism, and from genetic changes arising within the organism itself.

The means of magnification is the process that Darwin called natural selection. Through natural selection, variations that produce greater reproductive success in a specific environment tend to accumulate; there is a gradual drift in the direction of a population of organisms with a collection of characteristics that confer greater reproductive success in that environment.

This by itself would not ensure extreme diversity; two other factors enter here. One is the very large number of characteristics that nucleic acids and proteins have the potential to express. There are immense numbers of possible DNA sequences, coding for an almost equally large number of proteins. Furthermore, some possibilities for diversity exist even for one organism with a fixed DNA sequence, which can only produce a few tens of thousands of proteins. This is because proteins often act in groups—and the number of possible combinations is proportional to the number of distinct proteins raised to the power of the number of proteins acting together. This phenomenon seems to explain how large numbers of different antibodies can be produced by the immune systems of higher organisms.

The other factor that has led to diversity is the degree of variability of the environments in which living things have found themselves over the long history on this planet. The fact that there are many different environments on earth, and that these environments have changed over geologic time, implies that there were many potential combinations of protein and nucleic acids that, in some circumstances, would have led to increased reproductive effectiveness—and therefore new species. However, as species reproduce themselves, there can also be neutral variations, as well as those tending to greater or lesser reproductive success. It is by no means true that all of the different forms of life that have existed are the result of previous adaptations to specific environments. Such adaptations are simply one of the important factors leading to such diversity.

One question that biologists are only beginning to investigate is that of the mechanisms for evolution—i.e., what are the cellular structures that made it possible for great biological changes to take place over the course of earth's history. Biologists know *when* the most significant changes occurred. It is clear, for example, that a great increase in the range of living things occurred after the first eucaryotic organisms developed about 1.2 billion years ago. Probably a similar increase took place at a later time, about a billion years ago, when multicellular living things began. Yet the mechanisms that would explain these spectacular periods of biological change and plenitude are only partly understood.

One suggestion involves a recently discovered property of eucaryotic cells. The DNA in such cells contains introns, long segments that do not seem to be involved in either the production of proteins or the control of cellular processes. The American biologist Walter Gilbert has suggested that these introns facilitate recombinations of segments of the DNA to produce new combinations that code for proteins or that control such coding. The ability to do this more rapidly could give organisms an evolutionary advantage in periods of rapid environmental change.

However, it has not been definitely established that introns have this function, or any function at all. Not everything in DNA need be useful to existing organisms. DNA is the repository of random evolutionary events that have taken place over the whole history of life. Although organisms have methods for editing DNA to remove some inessential features, these editing mechanisms are not perfect. Biological evolution is not like an architectural process; it isn't like building a preplanned structure in which every element that is included plays some essential role. It is more like a process of tinkering, in which new elements are attached, while older elements that have become useless are not necessarily pruned away.

There is another theory that would help explain the diversity of eucaryotic cells. There is evidence that these cells are the result of a symbiotic association among several subcellular parts that, in an earlier stage of evolution, were freely living organisms. The original association might have been accidental, but it gave the organisms an evolutionary advantage. There was a kind of division of labor within these new cells. The originally independent organisms were now able to specialize in certain functions in such a way that the cell would thrive in environmental situations where the individual organisms could not have. The division of labor would also allow for a wider variety of cell types and activities. Because the necessary metabolic activities that every cell must perform were done more efficiently, there was more energy and time available to perform other activities. Eventually, the associated parts learned to synchronize their reproduction so that the cell as a whole reproduced itself. There is some observational data about present organisms that supports this view, including the existence of independently reproducing DNA in mitochondria, nonnuclear cell structures that play important roles in the metabolism of cellular energy. Definitive evidence for this theory of the origin of eucaryotes remains to be found.

Whatever the origin of eucaryotic cells, we know a few things about how they achieve their present diversity. Eucaryotic organisms typically have more information in their DNA than do organisms without cellular nuclei, such as bacteria—but they do not necessarily use all of it at once. Therefore, there must be control mechanisms, which allow part of the information to be used while the rest is kept in reserve. Although this also happens in bacteria, the control mechanisms in eucaryotes appear to be much more elaborate. The very segregation of the DNA into a nucleus introduces a further separation between the genetic information and the cellular machinery that translates it into protein. This separation allows for one form of control in eucaryotes. Messages from the DNA are first "transcribed" into corresponding messages in RNA within the nucleus. But only some of this RNA ever gets out of the nucleus to deliver its message to the ribosomes, where proteins are formed. This selectivity makes possible greater precision about which proteins interact with which others than is the case when all RNA is translated as in bacteria.

The capability of eucaryotic cells to control the expression of their large store of information allows distinct cells in multicellular organisms to specialize in different functions. It is this specialization on the cellular level that allows for the immense variety of activities of multicellular organisms, from the singing of nightingales to the mating of elephants to the thinking of humans. A multicellular organism is not a collection of trillions of cells, each of which is doing the same thing. Instead, such an organism contains hundreds of cell types, each with different functions. These functional differences are the result of chemical and physical differences among the cells. Since there is good evidence that the genetic content of all the cells in an organism is the same, the differences in function and the persistence of these differences when cells reproduce must depend on how the expression of this information is controlled.

Many of the things we would like to understand about living things are connected to the control mechanisms of eucaryote cells, including one of the major unsolved problems in biology: how organisms develop from immature forms to adults.

Living Things Are Part of a Biosphere

An important feature of living things is their interdependence. Most living things depend on other living things to survive. This dependence can be as obvious as the fact that animals must eat plants or other animals to live. It can also take on less evident forms, such as the compensating activities of two types of bacteria: one "fixes" atmospheric nitrogen into organic compounds, the other releases nitrogen from these same compounds to replenish the atmospheric supply. All higher organisms depend on others in this way. Even some types of bacteria that do not seem to directly require other organisms for their activities, depend on them indirectly. The environment in which these bacteria thrive is largely a product of the activities of living things. If all other life on earth were destroyed, this environment would rapidly change, and these bacteria would, in all likelihood, be annihilated too.

One way in which the interdependence of living things is expressed is by biological cycles. Each atom of the biologically important elements passes through a series of incarnations. An oxygen atom that is exhaled by a chipmunk today as part of a carbon dioxide molecule may be inhaled by a maple tree next year and converted through photosynthesis into part of a sugar molecule, which eventually will be consumed by John Smith in a hundred years [Figure 4].

In order to understand this idea, we must recognize that the atoms that take part in biological cycles may spend only a small part of the time incorporated into the bodies of individual living

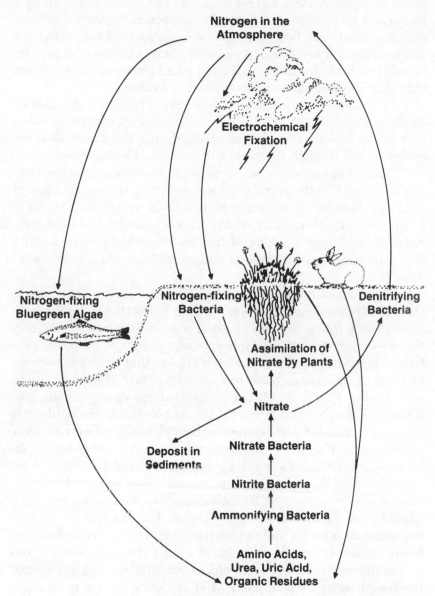

FIGURE 4   The Nitrogen Cycle in the Biosphere. The figure shows
some ways in which nitrogen flows from the atmosphere through
the soil and living things and back to the atmosphere.

things. The carbon dioxide molecule that the chipmunk exhales may spend years in the atmosphere before it is inhaled by the tree. A phosphorus atom that was once part of the body of a rhinoceros might lie in the African soil for centuries before it is taken up by a bacterium and included into its DNA backbone. Atoms may be exchanged between living things and other parts of the earth over and over again. Even when some atom is not in the body of a living thing, it still may be taking part in physical processes, such as rain, that play an important role in biological cycles.

In the nineteenth century, biologists introduced the idea of the biosphere, consisting of all of the matter on earth that takes part in the biological cycles and that, during the course of these cycles, passes through the bodies of individual living things.

Defined in this way, the biosphere is a very large system, consisting of much of the earth's surface, including the water, air, and soil. The amount of matter included in it is millions of times greater than that contained at any time within the bodies of individual living things. In spite of its size, the biosphere has a unity that is in many ways as great as that of any individual living organism.

The flow of matter from place to place in the biosphere is analogous to the flow of nutrients within living things. Just as individual living things need an external source of energy to maintain themselves, so does the biosphere—whose energy source is sunlight. There is a physical law which implies that an object with no external energy source tends to approach a state of equilibrium as time passes. This makes it necessary that the energy of the biosphere, as of individual living things, be constantly replenished, if they are to avoid the fate of death through equilibrium with their environment. The matter in the biosphere, on the other hand, can be recycled without the need for outside additions.

Although the idea of the biosphere was introduced to summarize the properties of biological cycles, some scientists have come to use the idea in a wider context. For example, evolution, expressed in terms of individual living things, or even classes of living things, is usually thought of as generating diversity, and producing new creatures from old. If we think instead in terms of the biosphere as a whole, then what evolution does is to increase the complexity of the biosphere, and enable it to include ever more of the material of the earth into its ongoing cycles.

The evolution of the biosphere has taken place over 3.5 billion years, and has resulted in a highly integrated system involving most of the surface of the earth and most of the matter on that surface. The increased diversity of individual living things that has taken place during this evolution can be thought of as a tactic that the biosphere uses to increase its complexity and range. These two ways of thinking about evolution are analogous to two ways in which we can think about the growth of an individual organism. On the one hand, growth involves the generation of a diverse population of different cells out of a single progenitor. On the other hand, the growth process results in the production of an individual organism of ever increasing complexity. From this point of view, the generation of the multiple cell types is a means, rather than an end. In recognizing the existence of the biosphere, people have realized that they are the conscious part of a vastly greater whole.

It became possible for the cells in a multicellular organism to specialize because other cells were available to perform some of the tasks that free living cells performed for themselves. Similarly, the fact that living things co-inhabit a biosphere allows many of them to specialize their activities, relying on other living things to carry out functions that they would otherwise have to perform for themselves. For example, few living things can directly use atmospheric nitrogen to produce amino acids. Most living things rely instead on the capacity of those bacteria that can do this; most living things obtain their nitrogen indirectly from these bacteria or from the plants that live, symbiotically, with them.

In the case of multicellular organisms, the individual cells with their specialized functions develop from a single cell toward the definite end result of producing a functioning adult organism. We have no reason to think that the biosphere is directed in a similar way. We do not know whether, if it were possible to repeat the course of evolution of life on earth, living things at all similar to those that presently exist would again evolve. It seems highly unlikely, because the course of evolution has been dependent on accidental environmental factors. Such an attempt to repeat history would probably yield a biosphere of some kind, but it would not produce the same specific individual manifestations. In this respect, the evolution of the biosphere is similar to the adult life of an individual, whose eventual form is much more dependent on accidental factors than is the development of an embryo from a fertilized egg.

*   *   *

Biologists have learned many things about life, on all its levels of organization. Nevertheless, it has been difficult for them to apply the general ideas that have emerged from their studies to produce a detailed understanding of many of the activities of living things. This is especially true for integrated activities of multicellular organisms, such as the control of behavior by the nervous system. The main reason for this difficulty is the degree of complexity of biological activity. Many elements are involved in even the simplest of living things, and their activities involve many interacting processes that go on simultaneously. Because of these interactions, even a complete understanding of each separate element of living things would not necessarily result in a complete understanding of all the activities of life. Even if we knew the function of each cell in the human body, we would not know everything about what the body does. (This is somewhat like the problem in understanding a novel in a foreign language by looking up each word in a dictionary.)

The intellectual problem faced by biologists is similar to although more extreme than the one faced by those physicists who are working to apply the general laws of physics to complex physical systems, such as solid bodies made up of many elements. In both biology and physics, new ways of thinking will probably be needed to find answers to the questions posed by the existence of complex systems.

Part 2
At the Threshold
of Understanding

Posing New Questions

The Quantum and
the Cosmos: Physicists
Search for Answers

Science often makes discoveries by finding answers to questions that have already been posed. By the late nineteenth century, many scientists recognized that ordinary matter was composed of atoms, but they had no convincing picture of what the atoms themselves were like. The discoveries of Ernest Rutherford, J. J. Thomson, and others at the end of the nineteenth and in the early twentieth centuries provided this picture. They showed that atoms contain electric charges, and that these charges are arranged somewhat like the planets and sun in a solar system.

Science will continue to advance by finding answers to old problems. But not all the unsolved questions of today's science will be answered. Some of them will be discarded, because scientists will recognize that they are meaningless, having grown out of an incorrect theory. This is what happened to an important question of nineteenth century physics, that of the structure of the ether, which was believed to transmit light waves. The question itself became obsolete after Einstein formulated his special relativity theory.

In some areas of science, new answers may bring us close to exhausting fundamental discoveries in the field. This has already happened in the field of acoustics, the study of sound. Though discoveries are still being made, they do not significantly change the overall picture of knowledge in the field. Other fields, such as cosmology, are more open-ended; it is unlikely that we will come close to the last word in the foreseeable future.

It is even less likely that we will arrive at the final answers to all of the questions of science at once. Even if scientists were to answer all the questions I will describe here, the answers themselves would raise new questions for future scientists. The only way to avoid this state of affairs would be to find a set of principles whose truth seemed so unquestionably apparent, it would preclude any further effort to account for the principles themselves. Science has not even attempted to go in this direction since the ancient Greeks, and I do not think it will do so now or in the future.

Furthermore, even if a field of science succeeds in answering all its own questions, there will remain the problem of how it relates to other sciences. If, for example, biologists were successful in providing biological explanations for all the phenomena of living things, there would remain the question of relating these explanations to the way in which physicists understand matter, living or not.

I do not agree with the biologist Gunther Stent, who suggests that because the subject matter of science is almost exhausted, progress in science may be near its end. Science will continue to face unanswered questions, both beyond its present frontiers, in cosmology and particle physics, and within those frontiers where existing explanations, in many cases, are far from providing detailed understanding of the phenomena. Individual areas of science may wax and wane, but as long as there are unsolved problems that can be attacked by the human mind and its computer extensions, it is unlikely that the whole scientific enterprise will end.

There are those who think that science cannot continue to make new discoveries indefinitely, since the human mind is itself finite. It is argued that there are aspects of the universe that we will never understand. Yet it is possible to accept the limitations of the human mind without concluding that scientific progress will end eventually. If there are matters that we will never understand, it is probably because they involve concepts and questions that will never occur to us. But it does not follow from this that we will ever run out of new questions to answer.

It is not hard for physicists to think of questions that they cannot now answer. There are questions involving what funda-

mental objects exist in the world, and how the numbers and properties of those objects have evolved in the past and will continue to
change in the future. There are questions about the general laws
that govern the behavior of these fundamental objects, and about
how to understand phenomena involving "ordinary matter" in
terms of the fundamental objects and laws.

Some of these questions are matters of fact, such as the precise number of subatomic particles and quantum fields that exist in
the universe. This type of question can be answered presumably
by the experimental methods that physicists have been using.
Other questions involve deducing the consequences of existing
theories, and so will require better methods of computation—or
even new mathematics to describe them. Still other questions will
require the invention of new theories to describe phenomena that
we have reason to believe are not contained within the scope of
existing theory.

Physics, as the science that deals with the most fundamental
phenomena, is largely on its own in finding answers to its questions, whereas sciences such as chemistry and biology can sometimes "borrow" the findings of other sciences. This view needs two
qualifications. One is that mathematics (not a science in my view,
since it does not have an empirical content) can act as an outside
influence on physics, and provide new ideas. Also, it sometimes
happens that physics can apply a type of explanation first used in
another science to its own very different phenomena. For example, we have already seen how physicists have used the idea of
evolution, originally a biological concept, to describe the changes
that occur in the subatomic particle content of the universe. Physicists may continue to get stimulating ideas from other sciences, if
not specific theories that explain the phenomena of physics.

Future physicists will also find themselves facing questions
that arise because of assumptions that are currently taken for
granted. I believe that all of the laws of physics, including those
that seem most obviously true, are ultimately factual and based on
observation rather than being "necessarily true" (as are the principles of mathematics). The physical laws that we know to be true,
are true in our world; but other logically consistent worlds can be
imagined in which they are false. We can, for example, imagine a
world in which total electric charge is not constant, but changes in
reactions among subatomic particles. Some of our current assumptions may turn out to be, if not false, at least incomplete.
Other assumptions may eventually be understood in terms of still

more fundamental ideas. Ideas that have been taken as obviously true by the physicists of one era have later come to be thought of as needing explanation; for example, the principle that all bodies fall with the same acceleration under gravity was further explained by the understanding that gravity is the result of curved space-time. Some assumptions, such as that space always satisfies the axioms of Euclidean geometry, have even turned out to be false.

While all of the assumptions built into present-day physics are candidates for explanation, modification, or replacement, some are more ripe than others. Certain ideas that underlie our description of space and time have already changed substantially in the twentieth century, due to relativity theory. Other parts of this description also have been brought into question—and in my opinion the description of space and time is the branch of physics likely to undergo the greatest changes.

The Final State of Matter—The Fate of Black Holes

One question we do not yet know how to answer involves the ultimate fate of matter that at some time becomes incorporated into one of the class of strange objects known as black holes. (This may actually involve much of the matter in the universe.)

According to Einstein's general relativity theory, any object with a large enough mass or a great enough density will distort space and time in its vicinity so severely that neither light nor any form of matter will be able to leave the area. Anything that comes near enough to the black hole will be unable to escape, no matter how hard it tries. The imaginary surface surrounding the black hole, which acts as a one-way filter such that nothing from inside the surface can escape, is called the black hole horizon. According to the theory, black holes might exist with any mass. This implies that the horizon radius, which is proportional to the mass, can also have any value [Figure 5].

One kind of black hole is thought to be produced by the death of large stars with masses several times greater than that of our sun (about 10^{33} grams). When such stars use up all their nuclear fuel, they collapse into black holes. Another type of black hole, which might be billions of times more massive than our sun, is thought to be formed as a result of the mutual attraction of many stars at the center of a galaxy. In each case the horizon radius would not be large by astronomical standards; the radius of those originating in stellar collapse would be only a few kilometers, while those at the

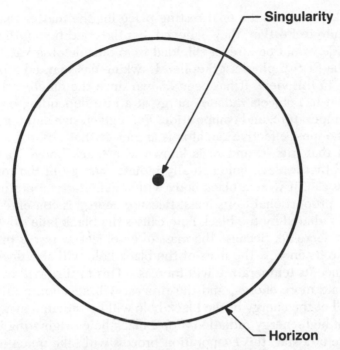

FIGURE 5   A Simple Black Hole. The horizon is an imaginary sphere through which only one-way trips are possible. In the region near the center of the sphere, called the singularity, we do not yet know what takes place.

center of galaxies would only be about the size of our solar system, a few billion kilometers.

Whatever the origin of a black hole, once it forms it is best thought of as a new type of matter, with specific properties of its own, among them being mass, electric charge, and spin. Any two black holes with the same value of these quantities should be imagined to be identical, even if they originated from different sources. It is reasonable to liken black holes to giant subatomic particles. Not only do the two objects have similar properties, but just as the properties of a newly created subatomic particle do not depend on the particles from whose collision it emerged, so a black hole, seen from the outside, gives no indication of the matter from whose collapse it emerged. However, the analogy between black holes and particles is not complete; gravity plays no significant role in the properties of subatomic particles, but it certainly does for black holes.

For some time, physicists thought that once a black hole was formed, it became the final resting place for the matter that had fallen into it. Further, they believed that the condition of the matter inside would be forever shielded from outside observation.

The British physicist Stephen Hawking has sparked a partial change in this view. It has been known since the nineteenth century that hot objects radiate energy at a rate depending on their size, temperature, and composition. For a given size and temperature, the most effective radiator is an object that absorbs any radiation that hits it, and so is known as a black body. Hawking showed that a black hole actually radiates energy in the form of photons as if it were a black body with a definite temperature inversely proportional to its mass. Because energy is conserved, the energy radiated by the black hole causes the black hole's own energy to decrease. Because the mass of an object at rest is proportional to its energy, the mass of the black hole will also decrease, and hence its temperature will increase. This in turn will cause it to radiate more energy, and the process will accelerate until almost all of the energy of the black hole will be radiated away in a burst of high-energy radiation. For a black hole originating in the collapse of a star, this evaporation process will take much longer than 10^{10} years, the present age of the universe, so it will not be easy to get direct evidence about the process.

There is no existing theory to describe what happens when, as a result of this process, the mass of a black hole has shrunk to about 10^{-5} grams—about the mass of a large amoeba. The radius of the black hole would only be about 10^{-33} centimeters at this point, and would bear the same ratio to the size of the amoeba as the size of the amoeba does to the whole observable universe. The present version of the general theory of relativity, which does not include the effects of quantum theory, implies that the radiation process would continue until all the mass has disappeared. At this point there would be no trace at the former site of the black hole of what has taken place, only a flow of radiation spreading outward through the universe.

This picture of the fate of black holes would imply that the process of formation and decay of black holes would result in a complete loss of all information about the objects that originally formed the black hole, such as what subatomic particles they contained, and how these particles were moving. Even though once the black hole is formed there would be no effective way for someone outside the horizon to find out about these objects, so long as the black hole itself exists, conditions inside the horizon might depend on their properties, so that the information about the original particles is stored away, rather than destroyed.

If the black hole itself disappears, any hope of retrieving this information would seem to disappear as well. However, it would be very surprising to many physicists if black hole evolution resulted in an irrevocable loss of information. The equations that describe this evolution do not appear to have any mechanism built into them for destroying information. These equations imply that if at some time, we know as much about a system as possible, then the system evolves so that we continue to know as much as possible about it. Nevertheless, because of the peculiar properties of space-time around a black hole, some physicists think that a loss of knowledge of the type described does occur.

An alternative view is that the photons and other radiation produced when the black hole evaporates contain as much information as the material that originally formed the black hole, and that this information could be recovered by trapping all the radiation and doing suitable measurements on it. This is what happens in a superficially similar process, the annihilation of a number of particles and antiparticles into radiation. This view, that black holes only change the form of information rather than destroy it, is a matter of controversy. More theoretical analysis is needed to decide if it is true.

Another possibility is that the black hole both stops radiating and stops contracting when it reaches a size of about 10^{-33} centimeters. If so, the end result of black hole evolution would be an object of ultramicroscopic size containing all of the information originally stored in a star or a galaxy. There is a reason why 10^{-33} centimeters is relevant. It is at this size that the effects of quantum theory on the evolution of a black hole are expected to become important. Unfortunately, we are not yet in a position to know definitely what those effects are.

A quantum theory of gravity has not been completely developed. There are several reasons for this lag. One is that until recently, there was little indication that the effects of quantum gravity would ever apply to anything we could observe, so that studying them seemed irrelevant to physics. Another reason is that the field theory describing gravity presents mathematical difficulties not present in other field theories. Finally, there is the peculiar influence of gravity itself on the properties of space and time.

Whenever we try to apply quantum theory to gravity, we must consider how these quantum effects will change the very arena in which the action is taking place, the space-time region surrounding the point of interest.

This last consideration is especially important at the very small distances mentioned above, because there the influences of quantum theory on the distortion of space-time take place over a region which is the same size as the object itself. Some physicists have reasoned that space-time, mass, and other familiar physical concepts change so radically under these conditions that these concepts are no longer a useful way to describe the world. Perhaps some of the more general abstractions invented by mathematicians may be needed instead. At present, we just do not know.

There is a related question about black holes that also seems to require a great theoretical advance. Suppose that a physicist, for his or her own information, were to take a one-way trip into a black hole to observe the phenomena inside (even though there would be no way to communicate this information to those left behind). If the black hole were galactic in mass or larger, this physicist would have a substantial period of time for observation before being sucked into oblivion.

According to the non-quantized version of general relativity theory, there is at least one point within the horizon of any black hole at which the effects of gravity become so large that the description of space and time at and near this point become meaningless. This point is referred to as a singularity. It has even been suggested that gravitational singularities may exist that are not hidden inside of horizons. If so, they could be studied from afar, through radiation that they emit and absorb. The problem of describing physical phenomena at a singularity is related to the problem of the end point of black hole evolution. Present physics cannot describe what happens in a small region about the singularity. The size of this region is about the same as the size of a black hole at the point when we no longer know how it evolves.

There is no convincing evidence that there are points in space-time, hidden or not, at which we will never be able to describe how physical properties behave. It would, of course, be necessary to modify and extend the laws of physics to make such a description of singularities possible. It is likely that the same principles that will enable us to understand the end state of black hole evolution will also describe the actual conditions at a putative singularity.

We do not yet know whether the necessary idea will be a quantum theory of gravity, or whether it will be provided by some unexpected phenomena occurring at a singularity or at the end of a black hole's life. These are urgent problems for theoretical scientists if we are to keep alive the sense that everything that exists in the universe can be studied by the methods of science. It is less certain that there is any prospect of observational or experimental input into these questions.

For black holes whose mass is originally of stellar or galactic size, the time that it takes them to evolve to their final state is so tremendously long that none could have done so since the universe began. However, it is possible that much smaller black holes were formed in the first instants of the universe, when densities and pressures were much higher than they are now. Such "mini" black holes might already have had enough time to evolve to a state near death, and it is possible that we could detect them from the high-energy radiation that they would be emitting by now. Attempts have been made to detect such radiation, but the evidence up to now has been equivocal. If we do detect it, a study of the radiation may tell us something about the final state of a black hole, and so resolve the problem of what happens to the information contained in matter.

Unsolved Problems About Space and Time

Over the past century, scientists have been engaged increasingly in the study of space and time. One result of this pursuit was the general theory of relativity, with its recognition that matter, through its gravity, can influence the behavior of space and time. However, there are questions about space and time that are not readily answered within this theory, as well as assumptions which the theory makes without adequate evidence. These questions, certain to occupy future scientists, are already being discussed.

1. Are Space and Time Grainy?

An essential element of the space-time description built into existing theories is that it is possible for subatomic events to be arbitrarily close together in either space or time. Another way to say this is that space and time are continuous (as are the unending decimal numbers) rather than discrete (separated by gaps like the whole numbers or like the atoms that make up matter). We do not have evidence that events can occur with arbitrarily small separation. What we have is experimental evidence that for separations in space as small as 10^{-16} centimeters, subatomic particles behave as if space and time were continuous, with no holes in between. Furthermore, some theories of subatomic particles have involved calculations that allow much smaller spatial separations than 10^{-16} centimeters. We can take the agreement of these calculations with observation as indirect evidence that if there are any gaps in space or time, the size of these gaps are smaller than the separations that are allowed in these calculations.

Nevertheless, some theoretical physicists have devised theories of subatomic particles in which space and time are endowed with a "grainy" structure. Most recently, such work has been done by the Chinese-American physicist Tsung-dao Lee and his collaborators.

In these theories, space and time are taken to be a discrete lattice of points, and motion takes place between these discrete points rather than continuously. The spacing of the lattice is taken to be small enough that the theory reproduces the known experimental results of the continuous theory. At the outset, the lattice was introduced for reasons of calculational simplicity. Calculations were being done in the branch of quantum field theory known as QCD. In some ways, the equations in this theory can more readily be approached using numerical methods than by using the more familiar analytic methods such as calculus.

Such numerical calculations are carried out by computers, which are better able to deal with motion in a discrete space than with motion in a continuous space. When the points in space are continuous, it is necessary to describe the motion at an infinite number of points, while in a discrete space, only a finite number of points are needed. The equations, in either case, were too complex to be solved without a computer, so the capabilities of the computer played a central role in deciding what approach to take. It was hoped that the computer's calculations in a discrete space, especially one in which the points were very close together, would yield a good approximation of what would happen in a continuous space.

Many such computer calculations have been done, and there is some indication that the results are similar to what physicists have guessed should be true in continuous space. This approach to QCD—usually known to physicists as lattice field theory because continuous space is replaced by a discrete lattice—was pioneered by the American physicist, Kenneth Wilson.

There is another reason, in the context of quantum field theory, for making space grainy. The equations used to describe subatomic particles in quantum field theories often yield mathematically absurd results that must be correctly interpreted to make sense. An important part of the interpretation process is to replace the original equations by a modified set of equations, which produce results that do make mathematical sense but which also depend on an extra number. In the modified theory, this extra number can have many different values, and the prediction that the theory makes for various physical quantities depends on the value of the extra number. At the end of the calculation, this additional number is given the specific value it would have had in the unmodified theory. This fixes the prediction for other physical quantities, and it is taken as the actual prediction that the theory makes for these quantities. This somewhat Byzantine procedure goes under the name of regularization. One way to interpret this regularization procedure is that the extra number is a minimum distance for the separation between two objects—here again the idea of "discrete" space becomes useful. At the end of the calculation, the minimum distance can be set equal to zero, and the equations become the same as those of the original theory.

renormalization

What was originally a concept taken up solely for pragmatic reasons later took on a life of its own. Some physicists have begun to ask whether we shouldn't take the idea of discrete space literally, as a bona fide theory of its own, rather than as a mere approximation scheme or a regularization method for a continuous space theory. This is an excellent example of how the means of calculation available to us influence our fundamental theories. And once a new idea leads to interesting results, it must be taken seriously whatever the original reason for considering it. The history of science is filled with examples of ideas whose ultimate significance is independent of the reasons for introducing them.

Various possible models for discrete space-time have been suggested, but it is too early to tell which of them is most plausible. One approach is to assume that there is a fixed underlying structure to space-time, somewhat like an actual lattice of points. According to this description, it would not make sense to ask what is "in-between" the points in space, because all motion or measurement would be limited to the points that actually exist. However, other questions would make sense. One is whether there are "connections" between the separate points that influence how motion takes place. That is, Can an object occupying one of a lattice of points more easily move to certain points on the lattice than to others? Such a question relates to what mathematicians call the topology of space and time. In this model, two other important questions are: How does the illusion of continuous space and time arise from the underlying lattice? and How is relativity theory (which is strongly wedded to the idea of continuous space and time) useful as even an approximate description?

An alternative to this idea of the fixed lattice would be that of a more "plastic" lattice. According to this model, although any specific example of motion is between a finite number of discrete points, the points themselves are not the same ones in different cases of motion. Any specific motion would take place between points of an irregular lattice [Figure 6]. When all possible motions are considered, every point in continuous space-time can be occupied. Probably this approach can be connected most easily with the usual continuous description. It also would not require an explanation for the fixed structure of the underlying lattice. It remains to be understood what new observable phenomena are implied by the plastic lattice theory.

Is such a radical change in our notion of space and time—from continuous to discrete—really necessary? I believe that it is, especially when we ponder the very short distances where quantum theory and gravity are important simultaneously. If this is the main criterion for determining the discreteness of space, then the distance at which this discreteness shows up would be the previously mentioned 10^{-33} centimeters. Perhaps when we are able to observe the final instants in the life of a black hole, we may get some insight into the ultimate atomicity, that of space-time itself. My own present work deals with this question. In the meantime, some physicists are actively pursuing the analysis of theories in which space-time has such a grainy structure. Such analysis may hint at new phenomena that can be looked for by experimental physicists.

2. How Many Dimensions Exist?

It is usually taken for granted that there are three dimensions of space and a single dimension of time. That is, any event that occurs anywhere in the universe can be assigned a location in space using three coordinates and a location in time using one. But physicists and mathematicians have studied hypothetical worlds in

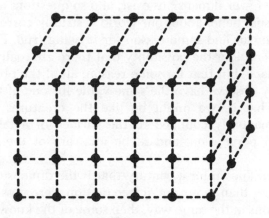

A. Regular Lattice

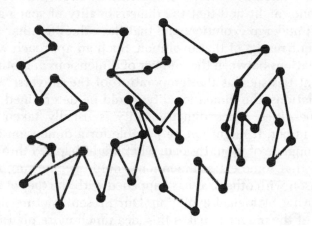

B. Irregular Lattice

FIGURE 6   Two Types of Three-Dimensional Lattices. In the regular lattice *A*, the points are arranged and connected in a simple way. In the irregular lattice *B*, the arrangement and connections of the points follow no simple pattern.

which more or fewer dimensions exist, and so questions arise as to whether the usual belief about our world is strictly correct, and if so, whether we can find some reason for it being true. For example, we might consider the possibility that there are really four dimensions of space, but that for some reason, all of the phenomena that we usually observe have the same value for one of the space coordinates. That is, we might be like the creatures living in Flatland, permanently confined to the surface of a table, both in motion and perception, and so be unaware of the extra dimension.

It has been known for a century that if the dimensionality of space were other than three, and if free motion were possible in all of the dimensions in the same way, then some of the known laws of physics would not obtain. Newton's inverse square law for the force of gravity is one such. This argument gives additional evidence that space is in fact three-dimensional, but does not explain why this is so. Furthermore, it does not rule out the possibility that our world has more than the expected number of dimensions, but that most phenomena are restricted in how they can vary in the extra dimension.

One approach to this question is to consider how spaces and times with different numbers of dimensions might behave. For example, one might find that the dimensionality of space and time can itself undergo evolution, and that the values familiar to us are the present results of that evolution. Such an approach would involve relations between the number of dimensions and other physical quantities such as the temperature of the universe. Through these relations, the dimensionality would be determined by these other quantities. Since dimensionality is usually taken to be a whole number, it might not be possible for a dimension to disappear through evolution. Instead, what might happen through evolution is that some extra dimensions could become suppressed in comparison with others, as has happened with the toes of some animals during biological evolution. Our present picture of the expansion of the universe makes this idea much more plausible than it once was. Since everything was once much closer together than it is now, we can imagine that there are indeed more dimensions than we think. The expansion of the universe may have taken place asymmetrically, so that in one of the dimensions there has been little or no expansion, and the scale of distances in that dimension would still be as small as it was at the beginning of the universe.

If this idea is correct, it would mean that there really are more than the familiar number of dimensions. It is intriguing to think that it might be possible to find some technological means to find and study the usually inaccessible dimensions. Very likely some phenomena would be different in a universe with more than four dimensions, even if there were no symmetry between the different dimensions. It would be of great interest to identify such phenomena and to see if they can be observed.

It is possible that these hypothetical extra dimensions are not only restricted in extent, but that they are connected differently than the familiar ones. Our space-time could be like the cylinder in Figure 7, open along its length but closed in the other direction. This would have interesting consequences for how physical quantities vary in these extra dimensions. For example, according to quantum theory, for every space dimension that is closed—such as a circle—there is an associated property—in this case momentum—whose value takes on discrete values, integer multiples of some unit. For the unrestricted coordinate of an open space the corresponding property can instead vary continuously.

Theoretical investigations have shown that if the general theory of relativity is set up in a space-time of more than four dimensions, and if the extents of the extra dimensions are made small and connected like a cylinder, then the resulting theory describes not only gravity, but also electromagnetism and other fields that have been introduced to describe subatomic particles. The extra dimensions in this case are associated not with space and time, but with the internal symmetries discussed in Chapter 1. Physicists are actively trying to unite space-time symmetries and internal symmetries in this way.

If other dimensions do exist, we would still want to account for the precise number, through some more basic principles. In the type of theory just described, the total number of dimensions would be related through an internal symmetry to the number of quantum fields that exist. But we should still need to understand why precisely four dimensions have expanded while the others remained small. The question of the dimensionality of space-time is ripe for more serious investigation. I expect that some new insights into it will emerge in the coming years.

3. *The Direction of Time*

Although the special theory of relativity is sometimes said to treat space and time on an equal footing, there are profound dif-

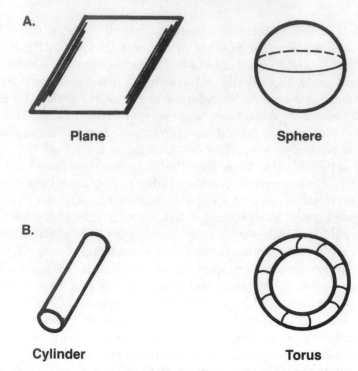

FIGURE 7   Two Types of Multidimensional Spaces. Each space is the surface of the figure shown. In *A*, each space has two dimensions that are similar. In *B*, the spaces have two dissimilar dimensions.

ferences between the observed properties of the two. While we are free to move through space at various rates, our motion through time is almost fixed. It is a consequence of relativity theories that the rate at which time passes can be different for two observers who are moving with respect to one another or who are in different locations in a gravitational field. Such effects have only been detectable by sensitive laboratory instruments, and have not shown up in ordinary human experience—because the relative motions of people have until now been very small. (The effects may become important in the somewhat distant future when we begin interstellar flight. At that time different people may be moving at relative velocities that are a significant fraction of the speed of light, and this is the condition under which the effects of relative motion on time would show up.)

However, there is a more profound difference between motion through space and through time, one that is unaffected by relativity theory. If we conceive of the instants of time as forming a line stretching into the past and into the future, then we move inexorably in one direction along this line, from past to future. The events of the past are accessible to us only through memory; those of the future are not yet accessible at all.

There is more to the directionality of time than just human perception. Many physical processes, especially those involving objects of greater than atomic size, take place as if these objects are being carried through time in a single direction. When two liquids, one hot and one cold, mix together, the combined liquid soon reaches a temperature between the two. The converse process—a liquid at a uniform temperature separating into two streams at different temperature—does not occur spontaneously in nature. Only an outside stimulus, such as human intervention, can make this happen.

But imagine a movie of the mixing process played backward to an unsuspecting audience. One would see on the screen precisely the reverse process to that observed in nature. If there are other processes taking place in the movie, such as the hands of a clock going round, the audience would realize immediately what was happening. An audience composed of physicists would not need such a cue to reach that conclusion.

By analyzing many such physical processes, physicists have concluded that there is a tendency for spontaneous events to take place in such a way that order decreases. The technical name for this conclusion is the law of entropy increase, and entropy is the name physicists give to one measure of the amount of disorder. In this context, a system is taken to be more orderly the more that it departs from an equal distribution among its possible states. For example, a number of atoms that were all in the same state would be a more orderly system than a mixture of the same atoms distributed among many states. In other words, order is identified with a deviation from what would happen if the atoms were assigned at random to the states. The law of entropy increase refers to spontaneous processes, meaning those that occur in an isolated physical system, unaffected by anything outside of itself. The notion of such isolated systems can only be taken as an approximation, because no system is ever completely isolated, but in many cases the approximation is a good one. If a system does interact significantly with its environment, then its order can easily increase, as has been the case for life on earth through its exposure to sunlight over billions of years.

There are basic questions connected with this principle of the decrease of order, and physicists have come to think of them as problems connected with the directionality of time. One question involves the contrast between what happens in processes involving large objects and those involving the subatomic particles or even slightly more complex systems such as atoms or small molecules. According to the fundamental laws of physics, there is no way to tell the direction of time in these latter cases. If we could make a movie of a subatomic or atomic process, there would be no way we could tell whether the movie were running forward or backward; it would look "right" either way. The microworld is a world of randomness in which past and future are indistinguishable. Physicists say that the fundamental laws for such microprocesses "satisfy time reversal symmetry." Why, physicists ask, does time flow in only one way in the world of everyday life, when it appears to flow in both directions in the microworld?

Of those questions in physics that have not yet been answered to everyone's satisfaction, this is one of the oldest. It was first raised in the latter part of the nineteenth century. While various answers have been proposed over the years, there is still a sense among many physicists, including myself, that the last word has not been said on the subject. The usual answer is that the notion that time has only one direction in the world of large-scale objects is only approximately true. It is overwhelmingly probable that things will proceed this way, but not absolutely certain.

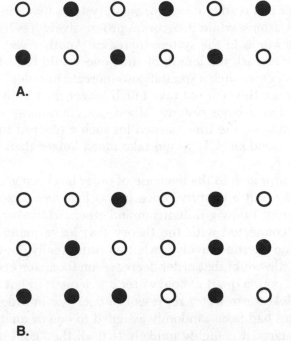

FIGURE 8   Regular and Irregular Arrangements. In *A*, two types of objects, represented by light and dark circles, are shown in a regular array. In *B*, the same objects are shown in one of a large number of possible irregular arrays.

It is argued that equally likely situations on the microlevel correspond to more disorderly situations on the level of familiar objects than to orderly ones. For example, a regular arrangement of atoms in a crystal lattice is just one of a very large number of possible arrangements, most of which are irregular [Figure 8]. Therefore, when changes in a system take place, such as the adjustments in temperature when two objects of unequal temperature are placed in contact, the changes are more likely to be in the direction of decreasing order. This decrease in order is what we mean by "entropy increase." It is the cornerstone of the physicist's concept of time moving forward in the macroworld.

According to this argument, it is possible to estimate quantitatively, for isolated systems of various sizes, the significance of possible deviations from the forward direction of time. For example, we can estimate how often a system will change spontaneously in the direction of increasing its order rather than decreasing it (a circumstance that might be interpreted as time flowing backward). This can readily occur for small systems involving no more than a few atoms, while it becomes progressively less likely as the number of atoms in the system increases. Another way to understand this is to ask the length of time one would have to observe the system before such a spontaneous increase in order occurs. For small systems, this will not take much longer than individual processes do. For a large system—albeit one containing as few as a thousand atoms—the time needed for such a reversal to occur increases tremendously. It would take much longer than the age of the universe.

This approach to the decrease of order has been generally accepted for about a century, since it was first recognized by Austrian physicist Ludwig Boltzmann and others. However, there are problems connected with the theory that leave many physicists uneasy. One question involves whether order really does decrease. There is little doubt that order decreases on the macroscopic level. When you pour a quart of cold water into a quart of hot water, the resulting lukewarm water represents a decrease in order because, if the atoms had been randomly assigned to one or another of the original quarts, it is highly unlikely that all the "hot" ones, those with higher energy, would have ended up in one of the quarts. But it has been suggested that on the microlevel, the amount of order always remains the same. We may think we are observing a process in which order decreases, but what is actually happening is that order is being transferred from one form, which is apparent to us, into other forms, that are not so apparent. In other words, the "decrease" is only in our information, what we know about the system, and not an objective change in the system itself.

While this idea appeals to some physicists, perhaps on the grounds that it is better if nothing is ever lost, other physicists resist it. They consider it an unnecessary introduction of subjectivity into physical laws. These points of view are strongly held by one or another group of physicists, and the fact that such disagreements persist is a strong indication that the question has not been completely answered.

It is possible that an experimental discovery will resolve this question, though it is highly unlikely. More likely is that there will be a new analysis of the problem, one sufficiently convincing to all future physicists. Such an analysis may also involve the way in which measurements on a physical system influence what we know about the system. This has been a controversial subject within quantum mechanics, and many scientists, including myself, believe that its resolution is connected with the question of whether and why order tends to decrease.

Since we observe macroprocesses to proceed from a more to a less ordered situation, we must ask how situations of high order can come about in the first place. As stated earlier, it is possible for a system that is interacting with an environment to become more rather than less orderly, as a result of energy input from the environment. For example, a pool of water with salt dissolved in it can change into water vapor and crystallized salt by absorbing sunlight. This does not contradict the law of entropy increase because the increased order of the pool is compensated by an equal or greater decrease in the order of the environment. In the present case, this is because the energy of the sunlight that is absorbed is more orderly than the energy that is eventually returned to the environment.

Through such processes, disorder can easily be transferred from one place to another in our universe, and in the process local concentrations of order can build up. However, this leads to the question of why the whole universe now appears to be in a relatively ordered state. For example, there are gross temperature differences between space and the stars. It must have been that at an early stage of the universe, a high degree of order somehow arose, which is gradually unraveling as time goes on. How the universe came to be highly ordered in the first place is a problem in cosmology, and eventually it must be answered in that context.

It is not difficult to see how some regions, such as the interiors of stars, whose density represents a high degree of order, could have emerged through a natural evolution of the universe. The expansion of the universe, which introduces an ongoing disequilibrium into the universe, plays a fundamental role in this way of thinking. Because the universe expands, matter that was in a state of maximum disorder at one stage of the universe became arranged into a state of order at a later stage, due to environmental influences such as the decrease in temperature of the surrounding space. Processes then took place that would have been impossible earlier. The type of matter found in the universe evolved, and most of the matter became consolidated into blazing stars moving through a cold, empty space.

The fact that space is mostly empty, and can serve as a repository into which radiation can be emitted indefinitely, is also closely connected to the expansion of the universe. This circumstance is very important for the appearance of order. If space were filled with high-temperature radiation, some processes that are observed to take place with the decrease of internal order would instead be replaced by a process in which internal order increases through the absorption of this radiation. Nevertheless, even in this "cosmological" approach to the decrease of order there remains a problem—for the universe appears, in its first instants, to have had much more order in it than it might have had. If there had been many black holes in the early universe, rather than many subatomic particles, the amount of order would have been as small as possible. Everything that has happened since the Big Bang has produced only a small decrease toward the theoretical minimum. The problem of the observed decrease in order therefore leads us to the question of how the universe originated.

What Happened Before the Big Bang?

Physicists describe the present universe as having emerged from a condition of immense temperatures and densities that were present about 15 billion years ago. It now appears that the evolution of the universe since then did not depend very much on conditions present at the origin. Physicists have gradually come to realize that many of the conditions in the present universe would have evolved out of a variety of initial circumstances. For example, we now believe that the surplus of electrons over positrons in the universe would have developed as it did whatever the relative numbers of these two particles at the beginning of the universe. Theories suggest that in the first instants of the universe, the relative numbers of electrons and positrons changed so that any difference in numbers was eliminated. The difference we now observe was produced slightly later, although still in the early instants of the universe. We know of other circumstances in which the end result is independent of the initial circumstances. An object of a particular weight and shape, falling through the atmosphere, quickly reaches the same terminal velocity whatever its initial speed.

From the physicist's point of view, this realization has both good and bad aspects. It means that we can hope to understand a great deal about how our world came to be the way it is without having to invoke arbitrary conditions that held at the beginning of time. On the other hand, it means that we are incapable of learning much about these starting conditions from anything that we can observe now. This hampers our ability to answer such questions as how the expansion of the universe got started, and what conditions were like before it began.

However, it is possible to use the methods of theoretical physics to look for answers to these questions, though one serious obstacle always gets in the way. It is the general relativity theory, which holds that the intense gravity present in the early universe would have distorted both space and time. Because of these important gravitational influences—and because of the possible role of quantum effects on gravity under these conditions (which we do not entirely understand), time itself may have behaved differently than it does in present conditions. Some scientists have concluded that it is not feasible to find laws that would have operated both before and after the beginning of our universe. As a result, they argue, we must accept the present universe as an entity unto itself, with its own peculiar laws, and that it is hopeless to imagine how the Big Bang itself might have evolved out of some earlier processes. For the most part however, physicists are constitutionally unwilling to accept arbitrary limitations on what is understandable. We continue to speculate about how the universe began.

Since we know that the universe has evolved on all levels since the Big Bang, it is plausible to think that evolution also occurred before the Big Bang. In other words, the Big Bang should not be regarded as the absolute beginning of the universe, but only as the beginning of the latest stage of universal history. Will earlier stages ever be amenable to scientific investigation? Though some scientists say no, this view seems as foolish to me as the nineteenth-century prediction by the French mathematician and philosopher Auguste Comte, that we would never know anything about the insides of stars. With this in mind, we can review some ideas that have already been proposed on the history of the universe before the Big Bang and on how the Big Bang happened.

One approach to the origin of the universe is based on the notion that the universe is finite and its expansion will not continue forever. The universe will behave somewhat like a rocket whose speed is less than that needed to escape from earth. Its expansion will gradually slow to a halt. This will be followed by a contraction, at first ever so slow but eventually speeding up, so that in the very distant future, the conditions present in the very early universe will be approximately repeated. Our "phase" of the universe will end in a Big Crunch.

According to the general theory of relativity, if the total density of matter in the universe is high enough, the universe is finite and this situation would indeed apply. Observations suggest that the actual density is at least 10 percent of the minimum value needed for a finite universe, but a better knowledge of possible nonluminous forms of matter in the universe is necessary before we can decide this question. (We believe, for example, that neutrinos—virtually massless particles with an electric charge of zero—exist in large numbers. We are less certain about the existence of other forms of "dark" matter.) If this scenario does describe what will happen in the future, then the same may have happened in the past, and our Big Bang could be the followup to an earlier period of contraction. In this model, one can imagine that the universe has existed forever, and has gone through infinite cycles of expansion and contraction, like a wire spring endlessly coiling and uncoiling and coiling again [Figure 9].

One problem with this picture involves understanding what happens at the midpoint of each cycle, when the expansion turns into a contraction. This expansion may be connected with the direction of time, and time may take on different properties as the universe reaches the point where expansion ends and contraction begins. In order to understand this, we need to conceive of time as something whose properties are plastic, rather than fixed, and then analyze how these properties would alter in the context of the change from expansion to contraction.

There are also problems in understanding the final stages of collapse (and the initial stages of expansion) of the universe, when quantum effects should become important, just as they are in the final stages of black holes. Since we do not yet know how to deal with these effects, we are in no position to determine how the change from contraction to expansion takes place—or whether it takes place at all. A proper quantum theory of gravity is becoming a necessity for understanding diverse aspects of the universe, including this one. The interest in these questions is likely to spur the formulation of such a theory soon.

Some physicists suggest that if the universe does indeed recycle itself, the amount of disorder in the universe must increase from cycle to cycle. If this is true, and if there have been an infinite number of cycles, then the present universe should be very

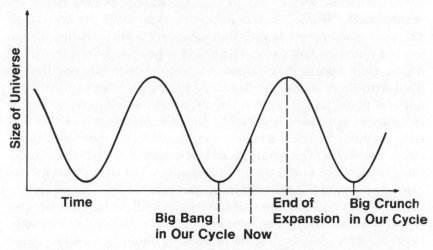

FIGURE 9   An Oscillating Universe. The relation between the size of a universe and time is shown for a universe whose size increases and decreases periodically.

disorderly indeed. As an analogy, picture a container filled with red and blue marbles, which are originally separated on two sides of a container. Now imagine that the marbles are poured into another container, then back into the first, and that this process repeats indefinitely. At each transfer, there will be some mixing of the marbles, so that after a number of transfers, their arrangement will be completely disorderly. So if disorder does increase from one cycle of the universe to the next, it would challenge the notion of an infinitely repeating universe, since in its early stages the present universe was much more orderly than it might have been if it contained many black holes.

I believe, however, that given the state of our present knowledge we don't know whether disorder really would increase from one cycle to the next. Furthermore, the problem of a relatively high degree of order in the early universe represents a general problem that must be solved, whatever model of the universe turns out to be correct. We cannot say yet that the oscillating model of the history of the universe is incorrect, but it is premature to accept it wholeheartedly.

Other scenarios for what took place before the Big Bang do not involve an endless sequence of expansions and contractions. One such idea relies on the possibility that "phase transitions" in the field content of space could have generated the high densities of subatomic particles that were present right after the Big Bang. The change in energy density that can result from a change in the level of quantum fields is so large that it can result in a transformation from a space devoid of subatomic particles into one that is filled with them. Thus, the Big Bang itself may have been the result of a phase transition in a universe previously empty of particles. If this approach is correct, then conditions before the Big Bang might actually have been much simpler than they were right afterward, just as the arrangement of atoms in a crystal is much simpler than that in the liquid that results when the crystal melts. This approach to the pre–Big Bang universe has its own problems, including the understanding of how and why the phase transitions would have occurred. The problem of the pre-history of our universe involves ideas both from cosmology and quantum field theory, and physicists are just beginning to investigate it. It will become an increasingly important question in future physics.

The Long-Term Future of the Universe

Scientists have given considerable thought to the origins of the universe, but until recently, few have thought very much about what will happen to the universe in the distant future. One reason they have shunned the question is that there is little hope of validating such speculations with any observational data. No human beings are likely to be present to test predictions about what will happen in 100 billion years. (Although we cannot directly observe the phenomena of the distant past either, we can often draw sound conclusions by observing their effects on the present.)

Nevertheless, there is beginning to emerge a scientific treatment of eschatology, the study of the late universe. I expect that eschatological analysis will progress rapidly and become an accepted part of theoretical science. It is only likely to become a part of observational science if it turns out that aspects of the universe are cyclic in time. If that is so, predictions about conditions in the distant future of our cycle of the universe could be tested by calculating the influence of these conditions on the next cycle, and comparing these predictions with what we know about the early stages of the present cycle.

Speculating about the distant future of the universe is an expression of how far the human mind can extend itself in space and time. Science originated, in part, with the aim of answering such questions, and that should continue even if no one will ever be able to confirm them. If one remains optimistic about our ability to modify undesirable aspects of the environment through intelligent effort, as I do, one might also argue that such speculation can illuminate and warn us about future catastrophes, which we can then work to avoid.

The long-term future of the universe includes both the overall structure of the universe, such as the distribution of matter into galaxies, and the properties of individual parts of the universe, from stars to subatomic particles. Since the visible universe is billions of light-years across, it follows that no significant change in the large-scale structure of the universe can be expected to occur over a time scale of less than billions of years. It *is* possible that changes in the properties of smaller objects, such as the "stable" subatomic particles that make up visible matter, could occur more quickly than this. After all, we know that such changes took place very quickly in the early universe. However, if we use current particle theory as a guide, it turns out that—with one possible exception, which I shall discuss later—changes in the subatomic particle content of the universe are also expected to occur only over very long time scales. For this reason, I will focus my attention on the very long-term future of the universe.

At present, we can identify two definite questions whose answers will profoundly influence our picture of the very long-term future of the universe. One of them is whether the universe is finite or infinite. The other question, which was mentioned in Chapter 1, is whether protons are ultimately unstable against change into less massive particles. I will discuss the future of the universe in three scenarios that take these and other questions into account.

Scenario 1: A Finite Small Universe

The universe can only be finite if there is enough matter inside it to curve space-time so that it closes on itself. We do not know whether there is enough matter to do this. We do know that the amount of visible matter is at least one-tenth as much as we calculate is necessary for this to happen—and that there may be enough matter that is presently invisible to make up the difference.

Some of this invisible matter may exist in the form of neutrinos, neutral particles that interact only very weakly. Neutrinos exist in the universe in large enough numbers so that if they have a small mass, they could provide enough energy density to close the universe. The same might be true about other, presently unknown, subatomic particles.

If there is enough matter present to make the universe finite, then there is also enough to cause the expansion to eventually stop and be replaced by a contraction. If this is the actual condition in our universe, then we would like to know when this changeover will occur. We cannot say this precisely, because of lack of information about the actual amount of matter in the universe, but we can say that it will not happen for a long time, probably at least as long as the time since the Big Bang—10 to 15 billion years.

It is usually assumed that the size of a finite universe would only be a few times larger than the size of that part we are presently aware of, about 10^{23} kilometers in radius. Yet there is no good reason for believing this, either on the basis of observation or of theoretical cosmology. If it were true, then the universe would begin contracting at a time in the future that is not much longer than it has already lived, so that our universe could be said to be middle-aged.

In this scenario, which I call the finite small universe, the future of the universe does not depend much on the details of particle physics. We know that nothing much can happen to change the properties and distribution of the subatomic particles in the universe over the next few tens of billions of years, until the density of matter becomes very high through the prolonged contraction of space-time. For example, we know that if protons are unstable, their lifetime is at least 10^{20} times greater than the present age of the universe, so that in the scenario under discussion, very few protons would have time to decay before the universe contracts back to the Big Crunch. Even the behavior of much larger constituents of the universe, such as many stars and galaxies, would remain pretty much as they are now during the remaining expansion time for the universe.

The study of particle physics as it relates to the future of the finite small universe is interesting only at two points: when the expansion is reversing to a collapse, and when the collapse reaches its final stages. These "eras" have been touched on earlier in this chapter.

Long before the collapse reaches its final stages, any of the large material structures, such as stars, planets, and their inhabitants, that exist in the current universe will have been destroyed by the increase in temperature and density that will take place during the period of contraction. Perhaps the most significant question about any finite universe is whether there is some way that intelligent beings could avoid being caught up in the eventual Crunch. Even optimistic writers, such as the American physicist Freeman Dyson, have asserted that this is probably hopeless. Yet I think that even in this scenario, the ultimate future for intelligence may not be completely bleak. The basis of my optimism is the notion that the space-time that we inhabit is not all there is, a view that a number of physicists have considered seriously, both for finite and infinite space-times. For example, it is intellectually irresistible to think of a finite "universe" as embedded in some larger universe, with a higher number of dimensions, just as the two-dimensional, finite surface of the earth lies in three-dimensional space. Indeed, the mathematical description of a finite universe makes use of such an embedding into a five-dimensional space-time. This approach to the idea of extra dimensions is different from the one discussed previously, because here the extra dimensions are not tiny in extent. Conceivably, both types of extra dimensions might exist. If there are "large" extra dimensions, we are free to speculate that other realms lie in this larger universe, and that their evolution need not parallel that of our own four-dimensional space-time. If so, we can then imagine finding some way of traveling between the different space-times, and avoid the fate of being crushed in the ultimate contraction of our own.

At present, this is no more than a science fiction plot. However, if there are more dimensions than those we know, or four-dimensional space-times in addition to the one we inhabit, then I think it very likely that there are physical phenomena that provide connections between them. It seems plausible that if intelligence persists in the universe, it will, in much less time than the many billions of years before the Big Crunch, find out whether there is anything to this speculation, and if so how to take advantage of it.

Scenario 2: A Large Finite Universe
From what we presently know about cosmology, it is possible that the universe is finite, but immensely larger than we can observe at this time. This possibility would require that the density

of energy in the universe be slightly larger than the amount needed to cause the universe to close.

If this is the actual situation, the universe will continue to expand for very much longer than it has already existed. Although it will eventually stop and begin to contract, our current space-time is now in its infancy. The long-term future of the universe would then depend on the behavior of matter, so that such a universe is of more interest to physicists than that of scenario 1.

In a large finite universe, very slow processes that could change the character of the matter in the universe would have time to act during the long period of expansion. Physicists have identified a number of such hypothetical processes, all of which act over time scales that are much greater than the present age of the universe, but which could be considered short compared to its ultimate life span. One such process is the aforementioned decay of protons into lighter particles, which, if it occurs, would require at least 10^{31} years on the average. Another possibility is that quantum mechanical effects will lead relatively small bits of matter to spontaneously collapse into black holes. It is difficult to make precise estimates of how much time such an event would take—but it is likely to take extremely long, much longer even than for protons to decay. Some matter will end up in black holes more quickly, as gravity makes some stars and galaxies collapse.

Although the ultimate fate of the black holes is unknown, as they slowly evaporate, most of the matter that was caught in them will be transformed into radiation by the process conceived of by Hawking. Therefore, whether or not isolated protons decay into lighter particles, if the universe continues to expand for long enough, much of the matter presently in it will ultimately be changed into photons and any other massless particles that may exist.

There is still another possibility, which is relevant if the universe contains large numbers of neutrinos and antineutrinos, or other weakly interacting particles with small mass. Like electrons and positrons, these neutrinos and antineutrinos can, when they collide, convert into photons, a process known as annihilation. Although the rate at which this happens is low, if the universe lives long enough, many neutrinos and antineutrinos will annihilate. There is a theory that the energy density of these weakly interacting particles produces the gravity that holds galaxies and clusters of galaxies together. If this is true, then their annihilation could lead to the instability of galaxies, the most characteristic objects in our present universe. It seems likely, then, that the most familiar objects in the present universe, from atoms through galactic clusters, are not eternal. They will disappear in the future, if the universe lives long enough.

This scenario should not surprise us. If the most important constituents of the present universe are destined to disappear, they will surely be replaced by something new. From what we know of the past evolution of the universe this has happened several times in the past, as the universe went from one dominated by many distinct particle species to one dominated by photons to one dominated by protons.

Some physicists have tried to describe the universe that would develop after the protons have decayed or the black holes have swallowed up matter as we know it. These considerations would apply either to the large finite universe of the present scenario (so long as it is still expanding) or to the next scenario, in which the universe is infinite and expands forever. The analysis is not complete, but it suggests that some forms of matter other than photons would persist in such a future universe.

The protons that are present in our universe would decay into positrons. These positrons can annihilate with the electrons already present to yield photons. However, the extent to which this happens depends on the rate of expansion of the universe, which, by increasing the average distance between particles, decreases the chance of annihilation. The analyses that have been given suggest that many of the positrons will find themselves too far away from an electron to annihilate. Consequently, some positrons, and an equal number of electrons, will remain indefinitely. The same appears to be true for neutrinos of finite mass, if there are any such particles. In any event, these particles that remain could form more complex stable structures, bound together by gravity or electromagnetism. These structures will be immensely larger than the familiar atoms; indeed, some may be larger than the present observable universe!

How complex these structures can become is an unsolved problem. It is difficult to analyze it in detail, because of the extreme disparity in scale between the structures that are familiar to us and anything that may evolve in the late universe. However, this change in scale is not unprecedented in the history of the universe. In its earliest moments, the whole region that eventually evolved into the present universe was much smaller than an atom, or even a subatomic particle. If there could have been an intelligence that functioned in the early instants of the universe, the familiar structures of our present universe would seem as grossly extended as the supergalactic atoms of the late universe would appear to us. It is not beyond our ability to understand complexity in the late universe, once we set our minds to it. I believe that understanding that complexity, and solving its related problems, will represent a novel branch of science in the future.

No matter how large, if the universe is finite, eventually the expansion will cease and contraction will take over. The details of what will happen during this contraction would be rather different from those in scenario 1, because the contents of the universe would be different in each case. Yet, the outcome is no less mysterious, so poorly are the phases of turnover and contraction understood. If we learn that the universe is finite, unraveling what will take place during these phases will become one of the important endeavors of future science.

Scenario 3: An Infinite Universe

If the density of matter is less than a critical amount, corresponding to about 10^{-29} grams per cubic centimeter—about ten milligrams in a region the size of the earth—then the universe is infinite, and will expand forever. Objects in the universe will, on the average, get farther and farther apart, except for those such as the contents of the solar system, which are held together by forces such as gravity. As in scenario 2, the contents of the universe will change, and their eventual form will depend on presently unknown properties of subatomic particles and on the end state of black holes. On the whole, the future of the universe in scenario 3 is about the same as in scenario 2, except that the expansion never slows to zero and reverses. The universe of scenario 3 becomes one in which photons, electrons, neutrinos and their antiparticles are spread ever more thinly through larger and larger regions of space. Once again, however, we do not know whether gravity and electromagnetism would allow these objects to form complex structures able to persist indefinitely.

However, there is a missing piece in our puzzle, one that might also apply to the large finite universe. Most cosmological models have assumed that the universe is homogeneous—that all parts of it are the same, including those beyond the reach of our telescopes and hence unknown. The assumption of homogeneity has been made in order to simplify the mathematical description that physicists give to the universe.

Recently, this assumption has been questioned. We have seen that physicists believe that some features of the present universe depend on the broken symmetry that occurred in the early universe. Yet this symmetry breaking need not have occurred uniformly over the whole of space-time. Just as a lake in winter can be liquid in some regions and solid in others, so might different regions of space be in different phases, which would imply different physical properties for the matter in it. For example, the surplus of what we call matter in our visible universe might be replaced by equal amounts of matter and antimatter, or a surplus of antimatter, in parts of space-time beyond our present horizon.

These considerations take a precise form in one specific cosmological model, recently proposed by the American physicist Alan Guth. In this model, known as the inflationary universe, all of the visible universe, and a great deal of space-time beyond it, originated in a tiny bubble in the very early universe. This bubble contained a high level of quantum fields, which caused it to undergo an expansion much more rapid than that expected according to the standard Big Bang theory, where the expansion is influenced only by the presence of subatomic particles. The rapid expansion diluted the material contents of our universe to a very low density, and none of the particles present in the universe before this "inflation" began are present today. Instead, the matter in the present universe was produced by a phase transition at the end of the inflation, in which some of the energy that was contained in background fields became converted to particles. In the inflationary universe, space-time beyond the volume of expansion of the original bubble would be different from space-time inside it. In this model, the different regions of the universe are somewhat like different cultures, developing independently of one another, and unaware of each other's existence.

We are presently unaware of conditions beyond our own bubble, because there has not yet been enough time since the beginning of the universe for any light originating outside to reach us. If the universe continues to exist indefinitely, we would eventually become aware of these foreign regions of space-time—and find that matter and energy in these regions have very different properties from those familiar to us.

But now let us return to the view that the properties of particles will change slowly as the universe expands. Some scientists have predicted that rapid phase transitions similar to those that took place in the early universe will occur in the future. This could happen if the present configuration of quantum fields in our region of space has more energy than another configuration, and is therefore unstable against transformation into the lower energy configuration—like a boulder teetering at the edge of a cliff.

If such phase transitions occur, they are expected to begin in one place, perhaps as the result of a random fluctuation, and then spread outward at the speed of light, eventually encompassing every point in space. As the transition passes through any point, those properties of matter that depend on the background quantum fields present there would have to change suddenly because of the new conditions. A sudden change in the properties of subatomic particles would lead to tremendous changes in any structures composed of them, and it is unlikely that these structures would persist. It has even been suggested that such a phase change has begun in another section of the universe, and is now approaching us at the speed of light. But there is no evidence for this possibility, and we need to analyze it further before adding it to the list of environmental catastrophes that we need to worry about.

If the universe continues to expand long enough for the matter within it to change its form drastically, then intelligent creatures may have a greater role to play in the distant future than they would in a universe that eventually contracts. They would have to grapple with two problems: the disappearance of the protons and bound neutrons that form the material basis for most structures in the present universe, and the ever smaller amounts of free energy that would be available to preserve order in whatever structures might replace them. No good solutions to these problems have yet emerged, but since we have been studying them for only a few years, and will not need the answers for 10^{30} years or so, we need not despair.

What Forms Does Matter Take?

Physicists think of quantum fields as the reality underlying our description of matter. This description is in many ways elegant, and works well with most of the things that we have observed about subatomic particles. However, in its present form, quantum field theory still has flaws.

An independent quantum field must be introduced into the mathematical description of matter for each of the known quarks and the six known leptons (particles such as electrons and neutrinos). Additional fields must be introduced to describe a number of other subatomic particles, such as gluons and photons, that interact with quarks. Even this large number of fields represents a substantial reduction from the view in the 1960s, when it was believed that there were hundreds of elementary particles, each associated with its own quantum field. But many physicists are uncomfortable with the remaining large number of distinct objects that presently occur in our mathematical description of nature. Furthermore, the equations that describe the behavior of the quantum fields contain many numerical constants such as the masses of particles whose values are not determined by the theory and so must be obtained from experiment. It has become clear to particle theorists that we have not yet achieved a completely acceptable theory of the structure of matter.

But we are trying. One idea is that quarks and leptons are themselves composite objects. If so, things would seem much less complex. The truly fundamental objects would be a smaller number of particles even simpler than quarks and leptons, and each quark or lepton would contain a few of these constituents. These constituents would be associated with quantum fields, and there would be fewer fields than in the present description.

It is not difficult to invent model theories of this type in which the individual fundamental particles bound together have very high mass but form composites with low mass. But such theories have not yet made any striking predictions. Should it turn out to be correct that quarks and leptons are composite objects, this development would simply be an extension of the process by which we understood how neutrons and protons were made up of quarks. One might believe that nature is more ingenious than this.

Another possibility exists within the framework of internal symmetry described in Chapter 1. It may be possible to regard the large number of different quarks, leptons, and other particles as aspects of the *same* object, related to each other by some overall symmetry group. This would not decrease the number of distinct quantum fields, but it would account for why that precise number of fields exists. It could also help us understand the precise numerical values of certain properties of the quantum fields and their associated particles.

Many theoretical physicists are engaged in this investigation.

The presently favored version of this inquiry involves relating fields associated with particles such as quarks to fields associated with particles such as photons—even though such fields differ considerably in their mathematical descriptions. The reason for the difference is that the particles such as photons, associated with one type of field, prefer to be found together, while those associated with the other type of field (quarks and electrons) prefer to shun each other. Furthermore, the spin of quarks is ½ unit, while that of photons is 1 unit; this also mandates a difference in the mathematical description. Nevertheless, the physicists Bruce Zumino and Julius Wess have shown that the fields associated with such disparate particles can be treated in a similar way.

The type of symmetry that can be imposed between fields that describe particles of different spin has come to be known as supersymmetry, and particle theorists are making active efforts to construct theories of the known subatomic particles that satisfy it. If this approach is successful, the theories that include supersymmetry will almost surely predict many fields and particles that have not yet been discovered, and so suggest new avenues of search for experimental physicists.

Even if we are able to find a complete description of the particles and fields that we know, together with whatever new fields and particles are implied by this description, there will be many surprises. The history of particle physics has shown repeatedly that subatomic particles are discovered in families that display a set of related phenomena. However, it is difficult to infer anything about the existence or properties of other families from what is known about one family. The behavior of quarks suggests that they occur in families, each of which contains two types of quark. For example, one family consists of the two types of quark that have survived into the present universe, and which neutrons and protons are made up of. From a study of objects containing primarily these two types of quark, it would have been almost impossible to learn about objects containing quarks from other families. The existence of other families was only discovered by producing these objects directly. A similar pattern has been found to be true of leptons. Because of the possibility that undiscovered families exist with masses too high for us to produce them in our laboratories at present, it would be rash to conclude that our current knowledge about particles and fields exhausts the phenomena of particle physics.

Even a successful theory of the particles we know would not guarantee that there are no other "islands" of phenomena waiting to be discovered. Perhaps we need to break through to some deeper level of understanding, in which quantum fields and particles are seen as manifestations of something more fundamental. With such a description, we might be able to show that only a certain set of objects with the general properties of quantum fields could exist; then we could be sure that we have discovered them all. This is not an impossible dream. It is similar to the present situation in nuclear physics, where on the basis of our understanding of nuclei as bound systems of protons and neutrons, we are convinced that we know which nuclei can exist as stable objects.

For the time being, however, we have no such underlying theory for quantum fields and their associated particles. The best we can do to determine which ones exist is to rely on empirical data. Some of the empirical arguments are quite subtle, and do not rely solely on direct observation. For example, many physicists are convinced that from an analysis of how helium and deuterium must have been synthesized in the early universe, it is possible to conclude that there are no more than four types of the particles known as neutrinos. If many more neutrino types existed when these elements were being formed, the extra neutrinos would have influenced the expansion of the early universe in such a way that more helium than is now observed would have been produced. A similar analysis, applied to the generation of an asymmetry between matter and antimatter in the early universe, might be able to tell us how many types of quantum field existed then.

All such analyses contain many assumptions and, in my opinion, cannot be conclusive. There is little choice but to continue the process of searching for new particles and fields under "novel" conditions, such as higher energy of colliding particles. In the past, our explorations of such new conditions have always produced surprises, and the greatest surprise of all would be if this pattern did not continue.

Unfortunately, the method for studying subatomic particles that has been followed for the past forty or so years is becoming extremely costly both in money and electric energy. The method involves inducing high-energy collisions between the particles and studying the resultant debris. In order to look for new particles with high mass and rest energy, it is necessary for the colliding particles that produce them to have even higher kinetic energy.

The production of large numbers of high-energy particles is done with one of several types of device known as a high-energy accelerator, which operates by exerting electric and magnetic forces on charged particles. As the desired energy increases, so does the size, cost, and energy requirements of the accelerator. The largest existing accelerators cost about $100 million and are about a kilometer in radius. Physicists are now talking of future accelerators that would cost several billion dollars and would have 10-kilometer radii [Figure 10]. Such accelerators would use hundreds of megawatts of power, as much as a city of a hundred thousand people. If such an accelerator is built, it would enable us to study particles with masses ten times larger than those we now know. This would be a welcome extension of our present knowledge, but it is hard to escape the view that this process is reaching its natural end, and that other methods for studying high-energy processes must be found.

Whatever our success in reducing or expanding the number of fundamental particles, or quantum fields, there are other problems in our description of matter that will require new approaches. Still further simplifications in our description of matter may be possible within the known equations of quantum field theory. Let me mention a few examples.

There is an important property of some subatomic particles, including electrons and quarks, called the Pauli exclusion principle. Invented in the 1920s, it states that no two electrons (nor two of any of the same type of particle with spin ½, ³⁄₂, etc.) can have exactly the same set of properties at the same time. For example, if two electrons have the same linear momentum, they must have opposite spin direction. It was believed for a long time that if a particle type satisfies the Pauli exclusion principle, then a specific mathematical property had to be assigned to the quantum field used to describe the particle. However, it has been discovered recently that equations written to describe fields corresponding to particles that do not satisfy the exclusion principle sometimes have additional solutions that behave like particles that do satisfy it. Perhaps this is also true for particles such as electrons and quarks. If so, we would be able to decrease the number of independent elements that enter into our descriptions because we would not need to introduce specific fields to describe these particles. This would be a radical approach to simplification, for not only would the number of elementary entities be reduced, but also

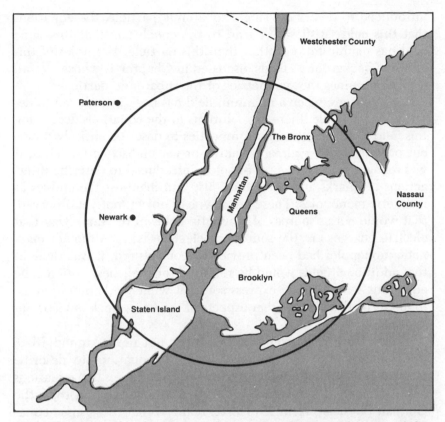

FIGURE 10 Scale of a Proposed New Accelerator. The size of the ring of a proposed new 10 Tev × 10 Tev particle acccelerator is indicated by the circle superimposed on a map of the New York metropolitan area.

the number of elementary properties. Similar ideas have been used to explain how the property of spin can emerge from a theory that does not include it at the outset. I believe that this sort of reduction in the number of elementary properties holds more promise for simplifying our description of the universe than does reduction in the number of elementary entities, and I expect that other examples of this trend will be found in the future.

The equations of quantum field theory may also contain unexpected information about objects that are not yet known. These equations may have solutions that describe entities whose existence was not suspected by the framers of the equations, nor by experimental scientists. Although a specific quantum field may be introduced to describe a single subatomic particle, the equations that this field satisfies may lead to the conclusion that this same field has manifestations other than this particle. Examples of this are the "background" fields discussed in Chapter 1, whose overall level determines the symmetries of the subatomic particles.

Sometimes, when a quantum field has been introduced to describe one particle, there are solutions to the equations describing this field that have the right properties to describe entirely different particles, either already known or yet undiscovered. Thus, it was recognized that one set of fields, introduced to describe interactions of quarks and of leptons, also can manifest themselves as magnetic monopoles. These are a type of object, not yet observed, that would act as sources of magnetic fields in the same way that electric charges are the sources of electric fields. Although magnetic monopoles had been previously conjectured, it was thought that additional fields would be required for their description. The equations we invent to express some of what we know often astound us by describing other aspects of the universe, known or unknown.

This feature of scientific thought is not new. Around 1870, James Clerk Maxwell framed a series of equations to describe electric and magnetic forces, and found that these same equations also described all the known properties of light. Around 1885, the German physicist, Heinrich Hertz, realized that those same equations implied the existence of what we now call radio waves, unknown before then. In 1927, the physicist P. A. M. Dirac invented a set of equations to describe spinning electrons, and found that they also predicted the existence of the electron's antiparticle, the positron, which was experimentally observed several years later. Perhaps physicists should not be surprised that our equations have wider manifestations than were expected by their devisers. These equations are often the result of unconscious thought processes and vague intuitions rather than purely logical constructs directly inferred from experiment. Because of this, the equations contain more than is needed just to describe known phenomena. This is also true of the equations of quantum field theory, and creatively working out the implications of these equations will be a powerful approach to obtaining new ideas in physics.

It is also possible that in the future, quantum field theory will not be the only approach to the description of matter. Quantum fields are only one type of mathematical structure that can be used to satisfy the constraints of quantum mechanics and special relativity theory. Some mathematical physicists have already identified other possible expressions consistent with these constraints. For example, structures called "strings" (because certain of their properties suggest vibrating strings) may be useful in describing families of subatomic particles whose description requires a number of distinct quantum fields. It is also possible that the relativistic quantum mechanics of these strings will avoid the problems of infinite quantities that beset the equations of quantum field theory.

Perhaps one of the alternative realizations of relativistic quantum mechanics will predict new phenomena. This is actually a more likely application of novel mathematical structures than is the replacement of quantum fields for describing subatomic particles. It would also be more exciting to scientists, who are usually more interested in studying new phenomena than in finding yet another explanation for old ones. For this reason, I believe that it is worthwhile to study the mathematical consequences of relativistic quantum mechanics, even when these consequences have no immediate application to the objects and phenomena that we know. This study can be one route that leads us to the discovery of what we do not yet know.

Finally, it is not only the structure of matter that may hold surprises for us in the future. One of the most interesting recent developments in theoretical physics was the discovery that space itself might have properties that would influence the properties of subatomic particles. We have not gone very far in exploring what properties space may have in the absence of matter, but I expect that a study of quantum field theory will suggest new possibilities.

Why Is There Symmetry in Nature?

Some years ago, a student who had done his doctoral thesis under my supervision was taking his final oral examination. At Columbia, this examination takes place before a panel of six faculty members, two of whom are not physicists and in this case, one of them was a prominent mathematician. Before it was his turn to ask questions, there had been some discussion among the panel members and the student about group theory, a branch of mathematics often used in contemporary particle physics. Then the mathematician asked the student, quite innocently, "Why is group theory applicable to physics at all?" The student evidently had never thought about this question. The mathematician withdrew the question as not relevant to the thesis and the student was awarded his degree. However, the incident led me to think about this question, and I concluded that the mathematician was asking something fundamental that requires an answer.

The use of group theory in physics is closely related to the notion of symmetry. We have seen that groups are mathematical structures used to describe situations in which there are symmetries among several objects, such as particles with related properties. The mathematician's question might then be rephrased: "Why should symmetries be relevant to our description of nature?"

A physicist might reply that the symmetry is actually in the phenomena that we observe and that our description simply reflects that fact. This is probably not the whole truth. Just as we have been led to ask why space is three-dimensional, we must ask whether the existence of symmetry in nature is fundamental or whether it is the consequence of something else. Some forms of this question have appeared in our previous discussion of how the apparent symmetry involved in relativity theory could be consistent with a grainy structure of space-time.

It is possible that symmetry is relevant only for phenomena of sufficiently large scale, such as those involving the known subatomic particles, but that on a much smaller scale there is no such symmetry. As an analogy, imagine a smoothly polished quartz sphere, one centimeter in diameter. If this sphere is rotated around an axis, its properties do not appear to change to the naked eye—a simple form of symmetry. But viewed through a microscope, the surface of the sphere will be seen to have many hills and valleys, and most rotations will lead to a different arrangement of these surface features, so that no exact symmetry exists under rotations [Figure 11]. The apparent symmetry occurs only because our senses cannot distinguish between those rotations that leave the surface unaffected, and those that change it. This is not the only possible explanation of how space-time symmetries can occur in nature; others have already been proposed and will continue to be studied in the future.

The question of symmetry becomes more pressing when we consider the internal symmetry groups physicists use to describe quantum fields and particles. As we have seen, it is not yet clear what the correct symmetry for this purpose is, so an obvious ques-

A.

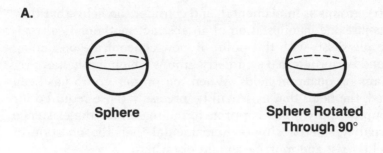

Sphere   Sphere Rotated
Through 90°

B.

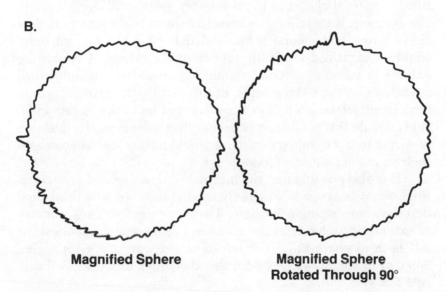

Magnified Sphere   Magnified Sphere
Rotated Through 90°

FIGURE 11   Symmetry of a Sphere. In *A*, a sphere is shown, along with the same sphere rotated through 90 degrees. At this scale of observation, no difference can be seen, and the sphere is said to be invariant under rotation. In *B*, the same sphere and rotated sphere are shown greatly magnified, and it is seen that rotational symmetry is not accurate on this scale, because of the many surface features that change position.

tion arises: Is any such symmetry relevant to nature at all, and if so why? Indeed, if we could answer these questions, it might lead us to the proper symmetry itself.

Scientists have pursued several approaches to the question of why internal symmetry groups exist. Many actually think of the symmetry groups as fundamental, and consider the fields that they relate as just one manifestation of an abstract mathematical reality. The physicists with this point of view concentrate on finding bigger and better internal symmetry groups which imply more relations among quantum fields. When the proper group has been identified, the fields that exist will be precisely those required for this group to apply. In this approach, the reason for the existence of internal symmetry is more fundamental than the question of what fields exist and must be sought elsewhere.

Other physicists take the existence of certain fields as the primary reality. They believe that if these primary fields can be identified, a natural choice of the symmetry group will suggest itself. For example, if there were reasons to believe that there were precisely three fundamental fields, and that all subatomic particles could be identified as manifestations of these fields, it would be natural to consider internal symmetries based on mathematical operations on these three fields. In this picture the existence of internal symmetries is a notion secondary to that of the fundamental fields. While this would not be a complete answer to the question of why internal symmetry exists, it would narrow the areas where such an answer could be found.

It is also possible that the internal symmetries and the space-time symmetries are more closely related than we now think, and that they have a common origin. There may be a single symmetry of natural phenomena on the subatomic level, which manifests itself both as space-time symmetries and as internal symmetries. Some of the higher-dimensional theories previously discussed suggest this very thing.

The notion of symmetry has been one of the most fruitful ideas in twentieth-century physics, and it will continue to be so. But we must always be prepared to question the sources of our scientific success as well as to examine the limits of their applicability. The study of the origins of symmetry in physics is just beginning, but I believe that it will become an important area of investigation in future theoretical physics. It may be one of the paths that lead us to knowledge of deeper levels of structure in nature.

The Molecule and the Biosphere: Biologists Ponder the Riddles of Life

Biologists have a fairly good understanding of the basic functions in unicellular life forms such as bacteria. This understanding relies on the biochemistry of the proteins and nucleic acids that play essential roles in all living things. Biologists also understand much of the behavior of larger organisms, such as how human beings digest their food.

However, there are wide topics within the general purview of biology in which little is understood. A serious unsolved problem is how the single-celled creatures came to be in the first place, that is, the question of the origin of life. For multicellular organisms, there is the unanswered question of how a single fertilized egg cell can become a complex organism. Related to this question of development is that of aging: Why do all multicelled creatures go through a process of senescence, ending in death? Though I will concentrate on these three questions, there are many other open questions in biology, probably more than in physics, and I want to briefly mention two others.

These two questions refer to two of the most complex aspects of highly developed organisms. One involves the immune system, a system of remarkable intricacy possessed by some multicellular organisms. This system gives a biochemical definition of individuality to each organism, and helps protect it from constant assaults from the environment. How the immune system operates, and how its operations are controlled by the genetic structure of an organism, are questions that are at the forefront of much contemporary biological research.

Another poorly understood part of complex organisms is the nervous system, which coordinates much of the interaction of each organism with its environment—and in humans provides another way in which each individual can be distinguished. Biologists are far from understanding in detail the operations of these systems, or how they can be understood in terms of cells and their interactions. These open questions, and the ones that I will discuss next in detail, leave a great deal for future biologists to do.

The Origin of Life on Earth

There has been life on earth for at least the last few billion years. Presumably there was a time, just after the earth was formed, when it was lifeless; the physical conditions then were such that the known forms of life could not function. Somehow, life developed out of this original lifelessness.

Many answers have been suggested to this question of the origin of life, both within and outside the framework of science. Scientists have even reproduced in the laboratory some of the chemical processes that may have been among the first steps along the road to life on earth. Nevertheless, none of the answers that have been proposed thus far come close to being satisfactory. Much more thought and research will be necessary before we know how life started. Oddly, although some prominent scientists, such as the British biologist Francis Crick, have written on the origin of life, there are not many scientists active in this area, in comparison with the number working on other fundamental problems.

The first individual living things on earth were undoubtedly very simple, probably even simpler than present-day bacteria and algae. We have no evidence that more complex multicellular organisms existed prior to about one billion years ago, whereas single-celled life dates to at least three billion years ago. But even the simplest of living things are extraordinarily complex by comparison with objects in the nonliving world.

One way to express the problem of the origin of life is to ask how something as complex as a bacterium could originate from nonliving matter. The same question can be asked about some of the individual parts of living things, such as the nucleic acid chains that carry hereditary information, or the proteins that catalyze the chemical processes in all known life forms. It is inconceivable that such complex molecular structures as proteins and nucleic acids could arise in a single chance encounter of simple molecules, even if such encounters were taking place in all the oceans of earth over a period of a billion years.

This does not mean, in spite of what some scientists, such as Fred Hoyle, have suggested, that one must seek for some divine or mystical explanation for the origin of life. One cannot accurately estimate the chances that some complex structure will be developed by assuming that this happens in a single step. Complex non-living structures on earth are not made in a single event starting from simpler structures. The Grand Canyon was not made in one stroke from a simple river bed. Instead, such structures develop over a period of time, as a result of many small steps that gradually point the process toward its end result. In such a process, the result of the earlier steps acts to constrain the later steps much more strongly than if these were taking place randomly. Raindrops hitting a surface may form a random pattern at first, but once part of the surface is worn away by earlier drops, the pattern may come to look far from random.

A plausible approach to the origin of life would be to try to identify the steps of a long chain of chemical reactions and physical processes, beginning with the simple molecules that might have been synthesized on the primitive earth and culminating in the kind of chemical complexity that characterizes the simplest known types of individual living things. In order to do this, we need to know more than the chemistry. It is likely that the early steps toward life on earth were strongly influenced by features that may have then been present in the environment: large bodies of water, which could have influenced the rate of chemical reactions; complex mineral deposits, which could have acted as selective catalysts; and variations in temperature and sunlight, which could have influenced which reactions would take place.

There is no general agreement among scientists about the environmental conditions on the early earth. Therefore, the problem of how life originated concerns complex chemistry in an environment of which we have little knowledge. Furthermore, we need a general analysis of the way in which order can be built up through a large number of simple steps. These are hard problems.

Several scientists, including myself and the American biochemist Robert Shapiro, have suggested that a primitive biosphere played an essential role in the steps that led to the emergence of life. These scientists have proposed that the types of chemical reactions and the cycles that different materials undergo in the present biosphere had precursors in the early history of earth, and that it was through the action of these primeval cycles over periods of time that the chemical complexity of present life developed. In this view, the individual living things are an expression of the biosphere, rather than the other way around, and the biosphere preceded individual living things in the development of life on earth. If so, then a useful strategy for studying the origin of life on earth or anywhere in the universe is to analyze how a biosphere may develop in various environments. This approach to the origin of life has only begun, but I expect that it will be one of the main directions in future research in this area.

Another question connected to the origin of life is whether there are substances other than proteins and nucleic acids that can act as catalysts and information repositories in living things. All the living things that we know use nucleic acids for the storage and transfer of information, and use proteins as chemical catalysts. Furthermore, there seems to be a universal code for the translation of the information into the synthesis of protein. But we do not know whether there are other substances that might play similar roles in environments other than earth, or whether there was a period in the history of life on earth when such other substances were biologically important.

In the evolution of living things, some species or larger classes of creatures exist for a time and then die out; only their fossil remains show us that they ever existed. Perhaps the same is true on a different scale; whole types of biochemistry may have played some role in early evolution, before the biochemistry based on protein and nucleic acid evolved. Conceivably, we might even find some "living fossils," microorganisms that even today use a biochemistry other than that based on nucleic acids and protein. Our knowledge of the present population of earth is far from sufficient to rule this out. A vigorous program to search for alternative biochemistries on the present earth would be a desirable way to extend our knowledge.

Unanswered questions concerning the origin of life abound even within present biochemistry. Most of the molecules found in living things occur in only one of the two stereochemical arrangements in which they can be synthesized in the laboratory [Figure 12]. We do not know if this is a result of the environment—an

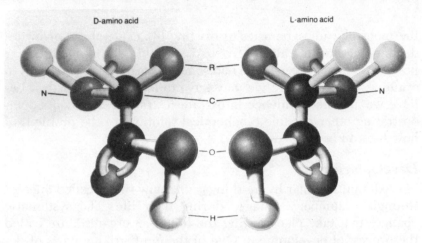

FIGURE 12   Stereoisomers. Two forms of amino acids, related by reflection in a mirror, can exist. All known biological amino acids occur in the L form.

accident of conditions on the early earth that has been handed down through the ages—or whether it is something intrinsic to the molecules, something that does not show up in simple experiments but becomes evident only over the course of a long evolutionary process. We also do not know why the present rules for translating a nucleic acid sequence into a protein sequence should be universal (or almost so). It might be that this involves some hidden chemical specificity—or again, it may be a result of the faithful transcription of an original "accident" in molecular evolution.

Finally, we need to understand not only how the chemistry of life evolved but also how some of the physical structures used by living things came about. It has been suggested by some scientists that the essential step toward the development of life came about when the first enclosed container, or primitive cell, was formed, allowing chemistry to proceed within it relatively undisturbed by the environment. These scientists suggest that what was inside such primitive cells was less important than the existence of the cells themselves, and that within the earliest cells, the biochemistry may have been very different from what it is today but still complex enough to be called life.

The present methods of biology and chemistry seem to be adequate to understand the molecular basis of existing life. But they rely heavily on the study of actual cases. It remains to be seen whether the same approaches will be successful in explaining how the molecular idiosyncrasies of existing life came about or in understanding possible alternatives. We would be in a much better position to find answers if we had some alternative life processes to examine—and for these we may have to look beyond earth. I believe we may find evidence in extraterrestrial environments demonstrating other possible biochemical solutions to the problem of how life can be carried on.

Development

All multicellular living things and many single-celled ones go through a number of stages during their life. The systematic changes that take place during the life of an organism are called the process of development. One of the most striking parts of development is its earliest stages. The life of multicelled organisms begins with a single cell, which contains the nucleic acid strand in which the chemical plan for the organism is written in a code that is the same for all organisms. But in multicelled organisms, in addition to specifying how the original single cell should function, the DNA also seems to describe how the single cell should multiply into what can eventually be many trillions of cells with a variety of different functions.

One question about development is the extent to which the DNA really determines the entire sequence of events that takes place in producing an adult organism. It is possible that the DNA in the fertilized egg cell contains detailed instructions for every step of development. Alternatively, it may only contain the plan for the first few steps, and what happens afterward may be determined by the result of these steps.

In either case, understanding the steps through which development takes place, almost unerringly, and with a different end result in each species, is a major problem of biology. Unlike the mechanism of heredity, which is now essentially understood in terms of the biochemistry of nucleic acid molecules, the mechanism for early development is a matter of great conjecture. Proposals have ranged from the conservative to the radical. Some speculate that development is yet another complex manifestation of the biochemistry of nucleic acid and protein, while others propose the unlikely possibility that it involves entirely new elements, not suspected from what we know elsewhere in nature, which satisfy different laws than other physical objects.

Though the mechanisms are poorly understood, biologists

have identified many of the individual steps in development and have recognized some of the molecular or cellular processes that are involved. Further work along these lines is continuing, and eventually we will know all of them. However, there are so many steps, each involving so many mechanisms, that even when we know all of the steps, an overall understanding of development will not yet be at hand.

Such an understanding would involve unraveling the relations between those elements of development that emerge from specific properties of the fertilized egg and those that take place through the action of general laws at various stages in development. It would involve understanding how a very large number of steps can occur in a coordinated way so as to result with high accuracy in the end product of early development, a baby organism that will eventually be able to function as an adult. It would also involve sorting out those aspects of the process that are essential from those that are not. There is no reason to think that every aspect of the development process is essential, any more than we should assume that every aspect of the structure or function of an organism serves a purpose. Some may have been important for some of the organism's ancestors and natural selection may not yet have had time or reason to eliminate them. (The human appendix is a well known example of such an evolutionary remnant.) Finally, an understanding of development should be tied in with some of the more puzzling aspects of the molecular biology of eucaryotic cells, such as the existence of large segments of the nucleic acid strand that, to our knowledge, do not code for proteins or other functioning elements of the cell.

Fusing many of the individual steps involved in development into a small number of processes that can be more easily comprehended is an important part of the biologist's quest. To gain some sense of what is needed, one might compare the steps in development to a computer program. Such a program may be described on a number of levels. The description may be as detailed as the individual changes that occur in the computer memory units, or it may be a less detailed but still complex description of the series of instructions that a programmer has written in a high-level language such as BASIC. On either level, a complete description of the individual steps may be available. But if the program is long and complicated there may be hundreds or thousands of steps, and such a description will usually not be intelligible to someone who has not written the program. We need something that gives an overall description of what the program does.

One form that such a description often takes is a flowchart, which describes the logic of the important steps of the program and how they fit together [Figure 13]. In a flowchart, large numbers of the individual steps are fused into a single large step, and the number of large steps is kept small enough for the human mind to comprehend. By looking at a program's flowchart, it is usually possible for another programmer to understand what the program does, and often to reproduce it. We need something analogous to the flowchart to describe the complex process of development.

In what follows I will describe some of what is presently known and not known, both about the individual processes that take place during development and the mechanisms by which they occur, and I will spotlight the progress that has been made in understanding development as a whole.

It is reasonable to consider early development as the first stage in the overall life cycle of an organism. Later stages include growth after the organism begins to function on its own, metamorphosis, in those organisms where this phenomenon occurs, and aging, the gradual loss of function with the passage of time. It is plausible that similar mechanisms act in all of the stages in development, perhaps distinguished by different time scales and by various environmental influences. Because early development (embryogenesis) and aging are the most pronounced examples of change, I will concentrate on them.

To understand embryogenesis, biologists must begin by clarifying the mechanisms that, in a period ranging from days to months in different species, transform a fertilized egg cell (zygote) into a mass of trillions of distinct cells containing in miniature most or all of the working parts of the adult organism. Several distinct changes occur during embryogenesis. Of these, the most straightforward is the increase in the number of cells, which occurs through cell division of a type not very different from that which takes place in the reproduction of simpler organisms. The two other major changes that take place during early development are cell differentiation, the production of the hundreds of different types of cells that occcur in the mature organism, and morphogenesis, the production of the multitude of shapes that are found in the embryo and in the mature organism. Both these processes are far more difficult to understand than simple cell division.

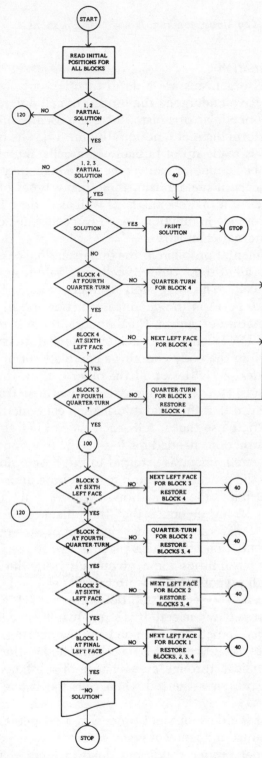

FIGURE 13   A Flowchart of the Program for the Computer Puzzle "Cubes." Each step in the flowchart represents many lines of the program.

Cell Differentiation

Complex organisms are able to function only because their components have undergone the process of cell differentiation. In a mature multicellular organism, there are usually many types of cells that perform distinct functions [Figure 14]. The human body, for example, is made up of human nerve cells, muscle cells, and white blood cells, among many others. Though they share some common aspects of metabolism, these many types of cells differ greatly in their internal chemistry as well as in their function. Indeed, the difference in function is largely a consequence of the difference in internal chemistry.

A fundamental problem in the understanding of cell differentiation is that all of these types of cells originate from a single fertilized egg cell. This egg cell contains more than enough information to perform the chemical functions of all the distinct cells in the mature organism. Furthermore, there is reason to believe that the DNA content of each of these mature cells is identical; every cell in the body contains a complete set of instructions for the chemical activities of all the other cells in the body. The question then is, How can so many cell types arise from the same source? How is it that the total information contained in the DNA becomes restricted so that it tells each type of cell to follow precisely the instructions needed for its specific functions? There are no guides in embryogenesis external to the developing organism. What guides the process is a combination of the information present in each cell and the interaction of one cell with another.

There is strong evidence that the influence of neighboring cells plays a crucial role in the development of each cell. The process of differentiation often takes place in two steps. The first step narrows the possibilities for a given cell from the wide range available to the zygote to a few alternatives available to each differentiated cell. The second step determines which alternative is actually expressed by the cell. It is strongly influenced by environmental factors, sometimes provided by neighboring cells, sometimes by more distant cells. It is not clear whether this influence is basically chemical, through an exchange of substances, or physical, through pressures exerted when cells touch one another, or both.

Much of the development process is a cooperative effort, in which the mutual influence of many objects that make up the system can, under proper conditions, lead to large changes in the

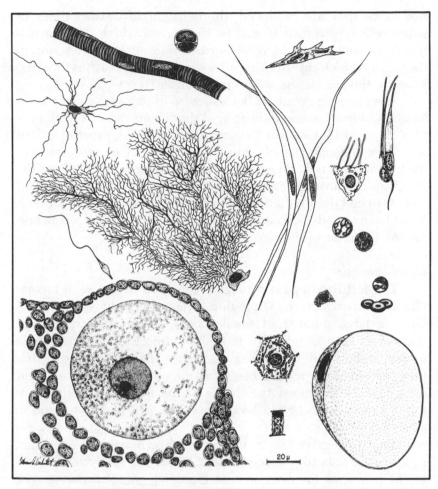

FIGURE 14   Cell Types. Some of the types of cells found in the human body, drawn to scale.

overall behavior of the system. Similar phenomena are known for simpler physical systems, such as a magnet or a laser. For some of these physical systems, it has been possible to give a mathematical description of this cooperative process; eventually it may be possible to do the same for some aspects of cell differentiation.

Meanwhile, progress has been made in understanding differentiation on the level of cell components. Even within an individual cell, a complex series of steps is involved. It is believed that a cell becomes committed to a specific differentiated state through a sequence of cell divisions that successively narrow its possibili-

121

ties. Once they are produced, the behavior of distinct differentiated cells is governed largely by the way in which the information in the nuclear DNA is transcribed into nuclear RNA, and in the way in which this RNA is processed before it reaches the ribosomes in the cytoplasm, where it is translated into proteins.

The complex details of the structure of chromatin—the combination of protein and nucleic acid that forms the cell nucleus—play an important role in transcription and RNA processing, but its precise influence is only dimly understood at present. Perhaps the invention of new experimental methods for learning about the structure of chromatin are necessary. Progress on the molecular aspects of cell differentiation is likely to continue rapidly, for biologists have already been able to answer similar questions concerning the behavior of simpler organisms.

Morphogenesis

The fertilized egg is a roughly spherical object with a complex internal structure. Through a series of cell divisions, it produces a large number of smaller cells, which at first arrange themselves very simply. But in a short time these cells, through motion and asymmetric growth, rearrange themselves into a complex pattern of shapes and begin to take on the form of the adult organism. Furthermore, the shapes of the individual cells themselves change into the wide variety shown in Figure 14 found in adult organisms.

This process, known as morphogenesis, remains a significant problem for scientific understanding. There are really several different problems involved in understanding morphogenesis. The problem of what mechanisms individual cells employ to move and change their shapes is the one that scientists know most about. Using electron microscopy, two sorts of filamentary structures, microfilaments and microtubules, have been identified in cells [Figure 15]. These structures, composed mainly of proteins, are attached to the cell surface and to other structures within it. They are able to exert forces that can alter the shapes of cells, inducing motions or permanent shape changes.

Another problem is how the shape of cells and cell components interact with the chemical events of cell differentiation. That is, how does the change in cellular biochemistry influence the actions of cell components such as microtubules to induce changes in shape? There is also the converse problem of how

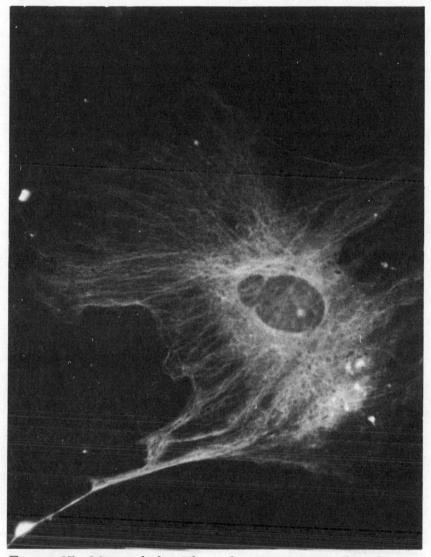

FIGURE 15 Microtubules. The embryonic mouse cell in this picture has been stained to display the filamentary pattern of microtubules.

the shape changes influence the genetic material in the cell to influence its biochemistry. There is reason to think that the interaction between biochemistry and physical form is a reciprocal one.

On a higher level of organization, there is the problem of how and under what influence groups of cells arrange themselves into the patterns that form organized tissue. And we want to know the nature and mechanisms of the influence of tissues on the shapes and chemical behaviors of the cells that they contain. It is on this level that the most serious gaps exist in our understanding of early development. A major open question is whether the information for this type of organization is coded within the genes of each organism, or whether this level of organization is the result of some type of "self-assembly"—by which I mean a mutual interaction of the cells once they have reached a certain state of development, and which is not very dependent on the detailed biochemistry of the cells. It may be that a number of cells of a specific type automatically arrange themselves into a definite pattern without requiring instructions to do so from an internal program.

There are arguments for and against such self-organization. Some observations about development are consistent with the notion that what happens to cells in one part of the developing organism depends mostly on the mutual influence of neighboring cells. For example, in some stages of embryogenesis cells can be removed from one part of the embryo, where they would ordinarily become part of the nervous system, and be transplanted to another part. The transplanted cells go on to develop into tissue of an entirely different type. On the other hand, it is not clear how self-organization can be specific enough that a mouse egg always becomes a mouse, and a bee egg always becomes a bee. This suggests that genetic specificity plays an important role even in the higher levels of organization, but leaves open the question of how important the role of self-organization is to morphogenesis.

There are many other questions that pertain to development on still higher levels of organization. We do not understand most of the steps that organize cells into tissues, tissues into organs, and organs into a functioning organism. We do not know whether there is a unique way for embryogenesis to take place or whether there exist alternative pathways for producing an adult organism of a specific type. Although substantial progress has been made, formidable problems remain. Understanding the process of embryogenesis is one of the major tasks of science.

Aging: The End of Development

Once an organism reaches the end of its growth, it remains on an apparent plateau for a long time, changing little in its overt functions. Yet even in this mature state, slow changes take place. There is a gradual deterioration in the biological functioning of the organism, both on the level of gross behavior and on the level of basic biochemistry. In humans and other mammals, this process of deterioration is what we know as aging. Eventually, this process results in the death of the organism, sometimes as the result of an inability to respond to an outside challenge, sometimes because of an internal event caused by the deterioration of some biological function.

It is not certain that aging is a universal phenomenon among complex organisms; it may take place only among those organisms that have internal limits on how large they can grow. However, there are also indications that something like aging occurs even in single-cell organisms. In any case, aging is widespread in nature, and it is a phenomenon that scientists would like to understand. Such an understanding will have immense effects on human life if it enables us to influence the course of aging.

Unfortunately, our present understanding of aging is even more primitive than our understanding of early development. We know neither the causes, nor in most cases the mechanisms, of aging. We do not know whether aging is intrinsic to development or an aberration of it. We do not know whether the fundamental process of aging occurs on the level of cells, on the level of the whole organism, or somewhere in between. There are tantalizing hints suggesting all of these possibilities, but none have been carried far enough to be convincing. There have been many theories of aging, but none have been able to synthesize the existing facts in any definitive way.

Scientists know much about the death rate and the related life span of many organisms. The death rate and life span vary a great deal from species to species. Furthermore, in humans and most other complex organisms, the death rate increases rapidly with chronological age. In humans over the age of thirty, for example, the death rate doubles every seven years. In other species, with shorter life spans, it doubles in a shorter length of time. Because of the rapid increase of the death rate with age, each species has a fairly well defined life span. Most organisms that die of disease, rather than accident, live to an age that is not very different from the average life span of their species. For example, about 70 percent of the Americans who died in 1981 did so within fifteen years on either side of the average life span of seventy-four years. By contrast, most laboratory mice die at about the average life span of their species, two years. Any successful theory of aging must explain the important facts that most members of any species live about the same length of time, and that this time varies so widely from species to species.

On the level of organ systems, there is convincing evidence that physiological functioning ebbs slowly, at a rate that depends on the species. In humans, the value of many measurable functions, such as breathing capacity, decreases by about a factor of two over fifty years [Figure 16]. The evidence suggests that all organs decline in function at a similar rate, but that organ function does not change nearly so quickly with age as does the death rate.

There is mixed evidence about changes in the function of individual cells over the lifetime of the organism. Some aspects of cellular metabolism seem to change considerably as the organism ages, others hardly at all. There is no accepted cellular property whose change over time can be used as a definite measure of aging in living organisms.

It has been possible to keep cells, and in some cases whole tissues, alive beyond the ordinary life span of the individual organism. One way of doing this is to lower the temperature of the cellular environment by so much that metabolism effectively stops. The cells can be thawed, and in many cases then go about their lives as if no time had passed during their frozen sleep. However, this is more of a way of preserving cells than of extending their life span. Another approach that has been used is to graft tissue, such as mouse breast tissue, from an old organism onto a younger one. In some cases this has been done sequentially several times, so that the original tissue survives for several life spans. However, this procedure eventually becomes ineffective. It is not clear whether this is because of some aging process that is going on in the transferred tissue or because of the trauma involved in the process of grafting. It is also not known whether *all* tissues have a potential life span that is substantially greater than that of the individual organisms they comprise, because only a few tissue types can be transferred in this way. Further experiments in this direction are needed to disentangle cellular aging from the aging of the organism.

Another experimental approach that is just becoming possible in mammals involves the transfer of parts of cells, such as the nucleus, from an aged organism to a cell taken from a young organism. Studies of how such hybrids function, in comparison with

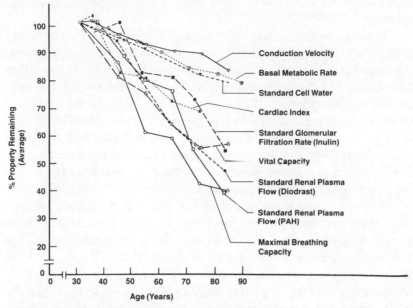

FIGURE 16   Change of Function with Age. The way in which several physiological functions decrease with age in humans is shown.

similar hybrids produced by transfer of nuclei from younger organisms, could give information about aging on the level of cell components and could also show whether aging involves changes in the hereditary material. The latter question could also be approached by determining the sequence of bases in nucleic acids obtained from aged organisms and comparing the sequence with that obtained from the same organism when it was young. This could be done both for the DNA in the nucleus and for the RNAs that act in protein synthesis, to see where, if anywhere, in the fundamental cellular metabolism the influence of aging is to be found.

Aging in the Test Tube?

Another source of information about aging comes from studies of cultured cells, those grown in artificial media. These media are combinations of those chemical substances known to be necessary for life. Cells originating in normal tissue are allowed to grow in plastic containers and are successively subcultured—that is, a small number are removed and transferred into new containers again and again to avoid overcrowding. An important discovery made by L. Hayflick and his collaborators is that when normal cells are grown in this way, they have a limited potential for dividing.

It has been found that two paths can be followed by the cells. If they are carefully screened to eliminate those that undergo chromosome changes, after a limited number of divisions—typically about fifty for cells derived from human embryos—the cells go into a state in which they can no longer be induced to divide. This transition is unrelated to the passage of time. If cells are chosen after a small number of divisions, maintained at low temperature or kept from dividing by some other method for a period, and then allowed to resume dividing again, they still undergo about the same total number of divisions.

There is some evidence that the maximum number of divisions that normal cells can undergo varies systematically from species to species, in a way that correlates with the normal life span of the species. There is also evidence that cells chosen from adults have the potential for fewer remaining divisions than do embryonic cells, and that the number of remaining divisions decreases with age. Some comparisons between cells that are near or past the stage in which they can no longer divide and cells that can still divide have revealed differences in biochemical functions, such as the production rate of certain enzymes. However, nothing has shown up in these studies that clearly explains the difference in reproductive potential in terms of such biochemical differences.

The other road that cultured cells can take is to undergo some chromosome alteration, which converts them into a form of cancer cell. Such transformed cells can grow and divide indefinitely. They are sometimes referred to as immortal, although strictly speaking it is the cell line, that is, a cell and its descendants, rather than the individual cells that deserve this name. A line of cells derived from Helen Lane, a cancer patient who died in the 1940s, is thriving in countless laboratories around the world; the cells will probably live as long as there are scientists to supply them with nutrients [Figure 17]. Evidently, by undergoing chromosomal transformations, cells can escape the limited growth potential of normal cells. There is some evidence that a simple gene replacement is responsible for this change from mortal to immortal cells.

Several other features of cell culture are of interest. One is that, in most cases, cells cannot be made to grow in purely artificial media. It is necessary to add serum derived from living organisms in order to get untransformed cells to grow. It was once

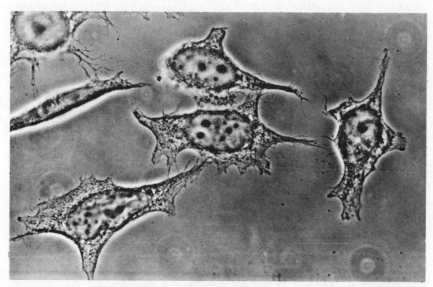

FIGURE 17   He La Cells. The descendants of cells originally taken from a cancerous tumor in a human patient. These cells apparently can divide and grow indefinitely, unlike normal cells.

hypothesized that the end of normal growth was due to the exhaustion of some vital factor in the serum. However, this was disproven by the fact that the same medium can contain both old cells at the end of their cycle of division and young cells that continue to divide. On the other hand, if certain substances, such as cortisone, are added to the culture medium, it is possible to increase the total number of divisions for normal cells by about 25 percent. Cortisone is a type of hormone known to form complexes that penetrate the cell nucleus and influence how information in the DNA is transferred to RNA, and it is this property that may affect division potential so dramatically.

Most cells in living organisms do not undergo anything near the number of divisions during the life span of the organism that can be induced in cell culture. Also, the conditions under which cells are forced to grow in the lab are quite different from those under which they grow normally. It is perhaps not surprising that under such artificial conditions, cells eventually cease to divide or transform into a new state for which unlimited division is preferred.

The fate of the cells that cease to divide is not well understood. Some experiments have shown that such cells can be maintained indefinitely in a state in which they continue to metabolize. This suggests that the loss of division potential is not necessarily the same as senescence. Indeed, it has been proposed that what eventually happens to the cultured cells that do not undergo chromosome changes is that they undergo a further type of differentiation from their parent cells, in which they lose their ability to divide, much as muscle and nerve cells do in living organisms.

It seems reasonable to regard the loss of division potential in cell culture as an important piece of information in the study of aging but not the crucial key to understanding this phenomenon.

Theories of Aging

However sparse the factual information about aging, the proposed explanations for it are legion. Although the number of scientists working on the subject of aging is not large, it often appears that each scientist working on aging feels impelled to provide his or her own theory, a pattern of investigation that is more characteristic of speculative philosophy than of science. It is not clear that any of the theories yet proposed are on the right track. Nevertheless, it is possible to categorize the proposed theories of aging and to see what types of experiments or further analysis they suggest.

Theories of aging need to address two fundamental questions. One is the cause of aging: What is it about complex organisms or their environment that makes them age at all? The other concerns the mechanisms of aging: What are the fundamental changes that take place in organisms as they age? These questions are related but logically distinct. There are many possible causes of aging that could lead to some of the same mechanisms for its operation. For example, either internally caused aging or environmentally induced aging might act primarily to damage a specific part of the organism, such as the immune system—and this damage might eventually result in all the other effects of aging. Conversely, a specific cause of aging might show itself by a number of distinct mechanisms, such as damaging one type of cell or causing a loss of function in all the different types of cell.

Since most aspects of the life of complex organisms are the result of a long evolutionary process, it is natural to ask whether there is some evolutionary function served by the aging process. The British biologist Peter Medawar suggests that aging has evolved to ensure reproductive effectiveness. According to his argument, some genes whose actions could have deleterious, even lethal, effects on an organism, can act at various ages in different members of some species. If the gene acts early in life, it is likely to result in the organism leaving few progeny. Therefore, if the time of action of the gene is itself an inherited characteristic, natural selection would tend to select those organisms for whom the action of the "dangerous" genes occur late in life, when the organisms have already reproduced.

Why then are these genes not eliminated entirely by natural selection? Medawar's answer involves looking at the population of the organisms that existed in natural conditions. Even in the absence of lethal genes, there would be deaths in this population unrelated to aging, due to accidents, predators, etc. As a result, there would be few chronologically old members of the population. Through this effect alone, most reproducing organisms would be relatively young. Since natural selection can only act on organisms that reproduce, there would be little or no tendency to eliminate deleterious genes that act only late in life. It follows from Medawar's analysis that the vitality of organisms should begin to decrease through the action of deleterious genes at about the age at which, in natural circumstances, accidents and other causes of death unrelated to aging would have destroyed most of the organisms anyway.

Therefore, according to Medawar, the aging process is the result of the genes whose action has been postponed beyond the average age of reproduction by natural selection. His analysis implies that the negative environmental factors that lead to death even in the absence of aging are important elements in setting the life span of some species. This suggests that those species least subject to predators, and with the most reliable food supply, would tend to have the longest innate life spans.

Medawar's theory of the origin of aging leaves several open questions. The argument does not suggest any specific mechanism for aging. Furthermore, the action of individual deleterious genes would not seem to result in the slow decrease in vitality that characterizes ordinary aging. Also, the timing of gene action itself is presumably part of development, and in Medawar's model it is unclear what sets the time scale for the action of the eventually lethal genes, and so determines the life span of each species.

Nevertheless, Medawar's theory probably explains in part

why aging still exists in spite of natural selection. It suggests that animals raised in laboratory conditions for many generations might gradually show an increase in life span, since the environmental conditions that fixed their original life span no longer operate. It also suggests a correlation between the length of the reproductive period in a species and its life span, a correlation that does seem to exist. This approach to aging would be more credible if we could identify the genetic components that have not been subject to natural selection. This could provide a mechanism to go with the suggested cause of aging.

Various mechanisms that operate on different organizational levels have been proposed for aging. In one class of theories, the primary events take place on a cellular or subcellular level and affect a wide population of cells of all types. These cellular events naturally influence the functioning of the body organs, and lead to the characteristic age-related changes observed in these organs, but these organ changes are secondary to the changes of the cells themselves. In another class of theories, the primary events are best thought of as occurring on the level of specific organs or systems of organs. These changes influence other organs only indirectly, but because of the interdependence of the body's organ systems, they eventually have deleterious effects everywhere.

Because complex organisms are highly integrated systems, it is not easy to determine where in the body or on what organizational level the primary events of aging occur. For example, suppose there is a biochemical change within liver cells, modifying the way in which they metabolize protein. Because the products of liver metabolism circulate in the bloodstream, they could soon affect the functioning of the brain, both on the cellular level and on the whole-organ level. The vulnerability induced by this systemic interdependence is the price that complex organisms must pay in order to function at all.

How can we tackle the difficulties that this interdependence presents to the study of aging? If we could continuously monitor the functioning of various types of cells and organs during the life of an organism, we might discover that the changes in organ systems that are associated with aging always take place in a definite order. Or we might find that there is always a specific biochemical or physical change within the cell that precedes other aging-associated changes. Such information would suggest specific mechanisms for aging. At present, we are not able to continuously monitor such changes within the body without drastically influencing the changes themselves. What is often done instead is to dissect animals that have died at various chronological ages and compare their cells and organ systems. This procedure is not nearly precise enough to obtain the required data. However, it may soon be possible to obtain such data by inserting microsensors into the body, which could remain implanted for long periods of time, and which would be small enough not to disturb the cell.

Strictly speaking, I should distinguish between cellular and subcellular mechanisms for aging. Eucaryotic cells are complex systems, containing many distinct interacting components. The complexity of eucaryotic cells is such that even if aging is basically a cellular phenomenon, it probably takes place differentially among the cell components. If this is so, futher questions arise. One is the problem of determining which of the cell components change during the aging process. For example, during the cell differentiation that takes place in embryogenesis, there is little change in the DNA sequence of the cells. Is the same true during the process of aging in the body? Is it true in the "differentiation process" that occurs during cell culture?

Evidence about this is becoming available through some of the techniques of molecular biology that have previously been applied to subcellular changes during embryogenesis. It has been found that in both the aging process as it occurs in the living body and in the process that takes place in normal cells when they are cultured, certain nucleic acid sequences are deleted from the DNA string, while others become amplified in it. The DNA strand in the aged cells is different from that in younger cells. Are these changes in nucleic acid an essential mechanism in aging? We are not yet sure, but such information is the first step in answering this question.

This question is further complicated by the fact that each component of a eucaryotic cell influences all of the others. Just as in the case of the age-related changes that occur on higher levels of organization, there is the problem of determining which changes in cell components are primary and which are secondary to changes in other cell components. There are prospective methods for doing this. For example, cell components can be transplanted to other cells, a method that will help us determine which changes in cell components are aging-related and which are the result of existing in diverse cellular environments. Then we will be in a better position to determine to what extent aging is a manifestation of such changes. Eventually, this type of information about aging on various levels of organization should enable us to find out whether there is a primary level on which aging occurs.

Another aspect of theories of aging cuts across the distinction between cellular and higher-level aging. It relates to whether aging should be regarded as a part of the normal process of development or as the result of presently unavoidable damage occurring over the course of normal life. Theories that take the former position are referred to as "programmed" theories while those that take the latter position are "wear-and-tear" theories. Certainly it is possible to imagine that aging is programmed to either operate on all cells or on the cells of a specific organ system. It is also possible to think that all cell types are worn down equally by living or that the wear is concentrated in a few essential types whose malfunction eventually damages the more resistant cells.

Theories that aging is programmed suggest the involvement of mechanisms similar to those operating in other stages of development, such as cell differentiation. This would suggest that progress in understanding one will help in understanding the other. However, since aging takes place over a much longer time scale than embryogenesis, other fundamental mechanisms may well be involved in it.

The wear-and-tear theories imply that aging involves mechanisms unrelated to those of early development and that some relatively simple factor may be at work to cause aging. These theories can be further divided into two groups: those that ascribe the agent producing the damage to something in the environment that the organism inhabits, and those that assume that the damage is produced by something in the process of life itself relatively independent of the environment. A prominent theory of aging that falls into the second subcategory has been proposed by British chemist Leslie Orgel. He suggests that the basic mechanism of aging is an accumulation of errors in cellular metabolism that originate in the process of protein synthesis within the cell. When a deficient enzymatic protein is produced by some accident, the action of this protein will cause a small disturbance in cell function, especially if the protein is itself involved in further protein synthesis. These disturbances can lead to the production of more deficient protein, leading to still more disturbances. The accumulation of errors within the cell continues, leading to an eventual loss of function. Aging is then identified with this loss of function over the whole organism.

There is some evidence that protein "errors" do accumulate with age in some cells, but the evidence is not uniform; nor is it clear that this accumulation is the result of Orgel's proposed mechanism. Also, there is no convincing argument linking the proposed loss of cellular function to the known facts of organismic aging, although perhaps that is too much to expect at this stage. In any case, Orgel's "error" theory is a fascinating one. Even if its proposed mechanism is not the cause of aging, it is worth investigating for what it can reveal about the processes of error and repair of molecules in cells. A variation of the wear and tear theories holds that aging takes place because the cellular repair mechanisms that are constantly at work to repair errors in nucleic acid and protein become less effective at a certain stage in the life of the organism. If this does occur, it might be the result of the "deleterious genes" postulated by Medawar's argument. Here too, definitive evidence for the effect has not been found.

There is at present no uniformly successful theory of aging. None of the mechanisms for aging that have been suggested thus far have successfully explained even the relatively few facts that are known, let alone predicted new results that have subsequently been verified. How we will understand the phenomenon remains unclear. Probably what is most needed now is not so much a correct theory of aging as a reasonable strategy for research, through which we could obtain more of the relevant information upon which a correct theory would be based. I believe that an essential element of such a research strategy is to study cells and cell components out of their normal environments. Such studies will enable us to separate the effects of aging that are intrinsic to these individual elements from those that are induced in them by their being part of an integrated system. They should also enable us to distinguish between environmental effects and those originating in the living process itself. This information should make it possible to formulate theories of aging that have a better basis in observation. We may hope that such theories will result in the understanding of what is presently one of the deepest puzzles about life.

All of the topics that I have discussed as puzzles for contemporary biology require the understanding of a complex sequence of events, involving many processes going on simultaneously and almost certainly containing elaborate feedback loops. I think it a safe prediction that future biology, and indeed future science, will need to develop ways of thinking about general processes of this type. We do not yet know whether there are universal features that describe all such complex parallel processes, but it would not be surprising if there were. Finding them would be a major scientific advance, comparable to the discovery that all types of motion could be described by the same laws. A good approach to this question may emerge from the experience we are gaining in designing computer programs, which share many features of these processes but which offer the advantage of being the result of our own design, and so more under our experimental control than those processes that occur naturally.

Beyond Microscope and Telescope: New Experimental Tools and New Discoveries

There have been discoveries in science, such as that of X-rays in 1895, that were not direct extensions of what was then known. Such novel discoveries, which can result in major transformations in science, are often made possible by new methods for experiment or observation. Two examples are the introduction of the telescope into astronomy and the microscope into biology, both of which took place in the seventeenth century. The telescope showed that the cosmos was immensely more complicated than had been suspected previously; the microscope revealed entirely new types of living things, as well as new structures (such as cells) in living things already known. A more recent invention is the radiotelescope, which has revealed many novel astronomical objects (such as quasars) and detected a sea of microwave electromagnetic radiation pervading all space—a remnant of the early days of the universe. Another revolutionary development is the high energy particle accelerator, which has revealed an immense number of distinct types of subatomic particles. The present quark theory of the structure of neutrons and protons is a direct outgrowth of this development.

Therefore, one way to anticipate novel discoveries is by analyzing which new methods for observation will soon be introduced. This is fairly easy to predict, because such methods are usually an outgrowth of existing science or technology. The use of new experimental techniques can also help to answer known questions in science. Novel discoveries are perhaps more exciting, but scientists would be very happy if some of our current problems could be solved through the introduction of new techniques of measurement. In this chapter I will discuss both answers to old questions and novel discoveries that may arise from new means of observation. I begin with some new techniques for studying cosmic phenomena, and then turn to techniques more relevant to laboratory phenomena.

Seeing Gravity

There is, at present, a group of remarkable experimental techniques under development for the observation of gravity waves. We are used to thinking of gravity as something that acts permanently between any two objects. But one consequence of the theories used to describe gravity, especially Einstein's general relativity theory, is that gravity has another aspect. The effects of gravity, distortions of space-time, actually become "detached" from the massive bodies on which they originate and travel through space at the speed of light until they meet other objects, whose motion they can then affect.

This phenomenon is similar to what happens with electromagnetism, which can take the form of light or radio waves. Gravity waves, like light waves, are thought to exist in many wavelengths, which will depend on the motions of the object producing the waves. The difference is that while light waves have been observed since there were human beings, and radio waves for the past century, we have not yet observed gravity waves. The effect of gravity waves on an object a few meters in size is much smaller than the corresponding effect of electromagnetic waves, and this makes detecting them much more difficult. Good detectors for electromagnetic waves, such as the human eye, can observe signals as small as 10^{-12} watts, the amount of light reaching the eye from the dimmest star visible to it. The best present detectors for gravity waves are only sensitive to signals of at least 10^{+9} watts, which for light would correspond to the signal produced by the most intense laser.

Furthermore, objects ordinarily produce many fewer gravity waves than electromagnetic waves. One reason is that most objects move slowly, and gravitational waves are effectively produced only by objects whose speed is close to that of light. Another reason is that gravitation is an intrinsically weaker force than electromagnetism. As a result, we must rely on cosmic sources of gravity waves in order to study them. It does not seem likely that within the foreseeable future we will be able to generate detectable gravity waves in our own laboratories the way that Heinrich Hertz generated radio waves in his.

Nevertheless, techniques are being developed that should be sensitive to the gravity waves that we believe are reaching the earth from sources such as neutron stars in formation. These techniques are based on the way in which gravity waves interact with matter, which is to produce tidelike forces in any object they encounter. Imagine four heavy masses arranged at the ends of the arms of a cross. A gravity wave that strikes these masses will cause the masses along one arm to move toward each other, shortening that arm, while the masses along the other arm will move apart, so that the arm lengthens. The actual distance that each mass moves will be extremely small, perhaps as small as 10^{-19} centimeter (much smaller than the diameter of an atomic nucleus) for a wave of a detectable magnitude.

Nevertheless, it should be possible to measure such small displacements in a variety of ways. One device, known as a laser interferometer, uses the fact that if light from an intense source is divided into two beams, and each beam travels a different path before the two are recombined, then the intensity of the combined beam shows a characteristic pattern of light and dark fringes depending on the difference in path length. In order to use a laser interferometer as a gravity wave detector, large masses are placed at the ends of two long crossed arms, which the laser light travels back and forth along [Figure 18]. When the masses are displaced by a passing gravity wave, the distance that the light travels along one arm increases, while the other distance decreases. This change results in an observable change in the pattern of light and dark fringes, which when measured can give information about the gravity wave.

No such effects have yet been detected in the prototype gravity wave detectors that have been built, but it is expected that, within the next decade, laser interferometers sensitive to gravity waves of an intensity of only 10^{-3} watts or less will be developed. This will be an improvement by a factor of a trillion over existing detectors. Other techniques for detecting gravity waves are also

FIGURE 18   Gravity Wave Detector. In this a prototype of gravitational wave detector that uses laser interferometry, laser beams pass back and forth through the long tubes shown, and are analyzed by detectors in the wire cage.

being investigated; one such measures the change in the light output of a laser when the region in which the light originates alters its size because of the effect of passing gravity waves.

There is an interesting sidelight to the quest for gravity waves. The changes they induce in objects of everyday size are so small that it is necessary to apply the laws of quantum theory to these objects in order to understand their responses. A prototype gravity detector now being built at Cal Tech contains long metal bars whose mass is about a ton. Suppose such a bar is hit by a gravity wave whose strength is within the detectable range. The change in the length of the bars that would be produced is about 10^{-19} centimeters. This is about the same as the amount by which quantum theory predicts that the length of the bar will be uncertain, because of the effects of Heisenberg's principle. This implies that measurements at this level of accuracy involving large-scale objects such as the gravity wave detector must take into account the principles of quantum theory that until now have only been needed when dealing with atomic or subatomic objects.

The detection of gravity waves would in itself be an important new piece of evidence for Einstein's general relativity theory. In addition, a study of cosmic objects emitting these waves should give us important new information about the universe. In general, the sorts of systems that produce gravity waves most copiously contain immensely dense parts that are moving at close to the speed of light. An example of such a system would be a pair of neutron stars orbiting one another at a distance only a little greater than their sizes. The orbital velocities would be sizable fractions of that of light. There is indirect evidence that one such pair of objects is indeed emitting gravity waves. The evidence comes not from the detection of the waves on earth, but from the effect on the stars' motion of the continuing loss of energy through emission of the waves. Other examples, yet unobserved, would be the last stages of collapse of a star into a black hole or the coalescence of two black holes.

We have no idea how many such objects there are in the universe, or even of the different types that may exist. Objects of this type are interesting to scientists for a number of reasons, one of which is that their description requires the detailed use of the general theory of relativity rather than just the approximate versions of it that have been adequate until now. Once we begin to systematically detect gravity waves, both our qualitative and quantitative knowledge about high-density objects in the universe will increase dramatically.

Because gravity waves interact so weakly with matter, they will pass right through any object other than a black hole. This is unlike the situation for electromagnetic waves, which can only travel short distances through matter before being absorbed, and so usually come to us from the outer surfaces of the objects that emit them. Gravitational waves can emanate from an object's core. If there are sudden rapid changes in motion going on deep inside of neutron stars, for instance, they can generate gravity waves, which, when detected, will yield information about the insides of such stars. (Unfortunately, even gravity waves cannot escape the pull of a black hole, so they cannot be used to learn about the inaccessible interiors of such objects.)

There is one other hypothetical source of gravity waves that may become accessible to us when sufficiently sensitive detectors are available. In some models of the early universe, the average energy of the particles in the universe was enormously high—so high that in the first few instants, gravity waves were being produced as copiously as electromagnetic waves. However, theoretical calculations imply that the rapid expansion of the early universe reduced the density of matter so quickly that the universe became transparent to gravity waves almost immediately. These gravity waves, remnants of the dawn of the universe, may have survived until the present, in numbers similar to the microwaves that have been observed with radiotelescopes. If so, the gravity waves from the early universe might be observable by detectors more sensitive than those now being built. If we can detect them, they might provide new information about conditions during the first instants of the universe. If we do not find them when we make detectors that would be sensitive to their expected magnitude, it may mean that some process has taken place between the time they were produced and now to eliminate these "fossil" gravity waves. This too, of course, would be worth knowing.

In a contemporary model of the very early universe known as the inflationary universe, an especially rapid expansion is held to have taken place in the first few instants, right after the early gravity waves were produced. This rapid expansion would have diluted the concentration of anything that existed before the expansion, including gravity waves, to levels so low that they would be unobservable in the present universe (which would not be the case if the early rapid expansion never took place). In this inflationary model, the objects such as subatomic particles and photons that exist in the present universe have all been produced after the early rapid expansion ended, when the more leisurely expansion that is still going on took over—by which time the temperature of the universe was too low to produce gravity waves in significant amounts. So finding fossil gravity waves in amounts similar to the amount of microwaves already observed would challenge the inflationary model of the early universe.

There are other means currently being developed for getting new information about astronomical objects. One involves the space telescope, an optical and infrared telescope that will be set up in space where it will be unaffected by the distorting influences of the earth's atmosphere. It will be able to resolve very small or very distant objects more effectively than earth-based telescopes. It will also be able to detect some objects that emit mostly infrared waves, which are largely invisible through the atmosphere. These might include the first definitive evidence for planets orbiting stars other than the sun. Even a planet as large as Jupiter orbiting another star would be very hard to detect through visible light because of the intense light produced by the star itself. Because it is possible to minimize the effect of gravity on orbiting telescopes, they can eventually be made much larger than earth-based ones too. Whether this new technology will lead to radical changes in our general picture of the astronomical universe remains to be seen. My own guess is that such changes are more likely to result from a new means of observation, such as gravity wave detectors.

Hunting Cosmic Fossils

A different approach to examining unknown contents of the universe involves looking for individual subatomic particles, rather than astronomical objects. The picture of the evolution of matter in the early universe that was presented in Chapter 1 suggests that the forms of matter that exist at present are only a small fraction of those that once existed. Most of the other forms disappeared long ago because they became unstable after a phase transition.

However, some yet undiscovered particles may have persisted into the present. In order for this to happen, these particles would have to have lifetimes longer than the present age of the universe. This could be the case if a particle has some property which is conserved (does not change with time) or nearly conserved. This is how protons have survived to the present. They carry a property called baryon number, which is nearly conserved, and none of the lighter particles that protons might decay into carry this property.

Such "fossil" particles might exist in small numbers. We are unable to reliably estimate how many would have survived from the early universe. Physicists have searched sporadically for specific types of such particles, such as magnetic monopoles and unbound quarks, but there has not yet been any systematic effort to find them.

A close relationship between particle physics and cosmology has been developing that is likely to stimulate a vigorous search for fossil particles. Until recently, particle physicists tended to regard the cosmological models as unworthy of experimental test, and they left such searches to the astrophysicists. Today the fossil hunt is thought to be a natural extension of laboratory particle physics, and so the two camps have joined forces. One reason for this is the realization that analysis of some aspects of the early universe can shed light on aspects of the present universe, such as the prevalence of matter over antimatter.

There is also a practical reason for the particle physicists' change of heart. High-energy accelerators have been used to study those types of particles, stable or unstable, that can exist with masses up to about fifty times the mass of a proton. For unknown fossil particles to exist today, their masses must be so large that we cannot produce them with present accelerators. Some theories predict the existence of subatomic particles whose masses are so high that we are unlikely ever to produce them (because of the immense amounts of energy and the very large costs involved). However, such particles could have been produced in the early universe and, if long-lived, still have survived to the present in small numbers. Such a fossil particle could only be found through methods of detection I will soon describe.

Heavy fossil particles—those whose mass is much greater than that of protons—are thought to be very rare in the universe as a whole, in comparison with the quarks and electrons that make up ordinary matter. This is known through an indirect argument. From observations of the rate of expansion of the universe, we have a reasonably accurate idea of the total energy density (the average amount of energy in a volume) of the universe. There are a number of protons and neutrons that we can detect directly, as they are part of visible stars or visible interstellar matter. The energy density of these visible protons and neutrons is at least 5 percent of the maximum total energy density that we know can exist. It is also possible that there are protons and neutrons that are not emitting light and consequently have not been detected (for example, those that comprise dark stars). It follows that the energy density due to any heavy fossil particles cannot be much greater than that of visible protons and neutrons. If a particle has a large mass, it also has a large rest energy, since rest energy is proportional to mass. Since the energy density of a particle is the product of its rest energy and its number density, it follows that if any individual particle species has a much greater rest energy than a proton, its number density must be correspondingly smaller. If a particle had a rest energy a thousand times greater than that of a proton, its number density could not be much more than one one-thousandth that of protons. Theoretical arguments also suggest that the survival rate against annihilation in the early universe of very heavy particles would be quite low, explaining their relative scarcity in the present universe.

In order to hunt for these heavy particles, physicists have returned to an old practice. Until about 1950, when experiments at particle accelerators began to reveal the rich details of quark-bound states, most discoveries of new particles came from a study of cosmic radiation, the stream of energetic charged particles that pervade the universe and constantly bombard the earth. These cosmic ray experiments found new particles that were secondary products of the collisions that the cosmic ray protons had had with atoms in the earth's atmosphere. Such experiments are likely to be less important in the future, because even the most energetic cosmic ray particles, colliding with atoms at rest, cannot create new particles with masses more than ten thousand times greater than that of protons. We will soon be able to produce such particles more reliably in the lab. Instead, the new experiments will concentrate on searching for rare charged particles in the primary cosmic radiation.

In order to get an idea of what would be involved in this kind of search, let us imagine that some fossil particle exists, with a rest energy that is a billion times greater than that of a proton but with the same electric charge as a proton. Some theories of subatomic particles suggest that such particles actually may exist. If such a particle type does exist in the present universe, then its total number must be smaller than the number of protons by a factor of a billion or more. Otherwise, the energy density stored in these particles would be greater than is consistent with observations of how fast the universe is expanding. It is plausible that if this heavy charged particle were to be found in cosmic radiation, it would be seen with a frequency of less than once for every billion protons counted. Concretely, this means that about ten such particles would hit each square kilometer of the earth's surface each second.

It is possible that some processes have occurred that increased the concentration of these particles in our part of the universe, compared to their average density. If this were the case, we would detect these particles in larger numbers—but we cannot afford to design experiments that rely heavily on this possibility. The small numbers that we can expect imply that if we are to search for such particles and eventually study their properties, we will need to use very large detecting systems or plan experiments that last for many years. Some such experiments have already begun.

The simplest search would be for particles with properties similar to those of known charged particles, but much heavier. Such particles, unless they are moving very slowly, would have a great deal of kinetic energy because of their large mass. They would not lose much of this energy through collisions, and would be easily detected in the standard detectors used for charged particles. These detectors, modern descendants of Geiger counters, operate through the effects of electrons that are ejected from atoms when a rapidly moving charge passes nearby. If the heavy charge can ionize atoms in this way, then observing them would just be a matter of setting up many detectors spread over a large area and waiting for one of them to show an ionization signal due to the heavy charge. This signal might be an excessively high amount of ionization, compared to that produced by a low-mass particle of the same kinetic energy. Such arrays of detectors might cover many kilometers and remain in place for years. These experiments could be automated easily, and checked periodically.

If, however, the charged particle is moving at less than about one ten-thousandth of the speed of light, it becomes more difficult to detect. Under these circumstances it cannot eject electrons from atoms easily. The detection of slowly moving charged particles is a problem that remains to be solved, though it should be possible to devise effective methods. But this probably won't be necessary. There is good reason to believe that if there are massive charged particles, they would not move this slowly. The gravitational force exerted by our galaxy on any such particle moving through interstellar space should accelerate it to speeds typically at least ten times greater. When the particles happen to encounter the earth during their wanderings, they are likely to have at least this velocity relative to the earth, and so be observable through the ionization that they produce.

Since most subatomic particle detectors "operate" by responding to the ionization caused by the particle's electric charge, different methods must be used to detect electrically neutral fossil particles. Some neutral particles might have strong nonelectromagnetic interactions with ordinary atoms. If so, they would have substantial probabilities of giving a signal while passing through a detector, providing that this detector is a few meters thick. These signals would be the result not of ionization, but of collisions with the atomic nucleus. Neutrons can be detected in this way. Such strongly interacting neutral particles could then be searched for with arrays of detectors like those described above, the only difference being the specific way in which the neutral particle interacts within the detector.

There is another possible source of the signal that would be produced by the types of collision just described: large fluxes of high-energy neutrinos. Neutrinos are a generic name for neutral particles of spin ½, without strong interactions; three types are presently known, all of which have very small or zero mass. However, some theories predict the existence of neutrinos with very large mass.

The known types of neutrinos exist as fossil remnants from the early universe. They can also be produced by collisions of cosmic ray particles with atoms in the earth. Even neutrinos produced in the laboratory are difficult to detect because each neutrino has only a small probability of interacting with the atoms in a detector. Because of this small interaction probability, neutrinos are said to have only weak interactions. Even for neutrinos of very high energy, for which the probability of interaction is greatest, a detector would have to be hundreds of thousands of kilometers thick in order to detect each neutrino that impinges on it. Detectors of this size are much too large to construct, and only a small fraction of the neutrinos that would pass through a detector of reasonable size would leave any record of their passage. As a result, significant numbers of neutrinos can be observed only if the number of them that hits the detector is large. For example, if the number of some type of neutrino hitting the earth were as great as the number of the primary cosmic ray protons, and the interaction probability were as small as expected, then a detector containing a million tons of material would detect about one neutrino every second whereas it would detect millions of charged particles. For smaller numbers of neutrinos—or with smaller detectors—the expected detection rate would decrease correspondingly.

Some physicists have considered the ingenious idea of adapting natural features, such as the oceans, as detectors for high-energy neutrinos. In this approach, all of the atoms in a very large amount of matter would be available as a target for the neutrinos. This large number could compensate for the small probability that the neutrino will interact with any individual atom. One project, known as DUMAND, is trying to use about a cubic kilometer (one billion tons of water) of the Pacific Ocean as a neutrino detector [Figure 19]. The idea is that high-energy neutrinos hitting the

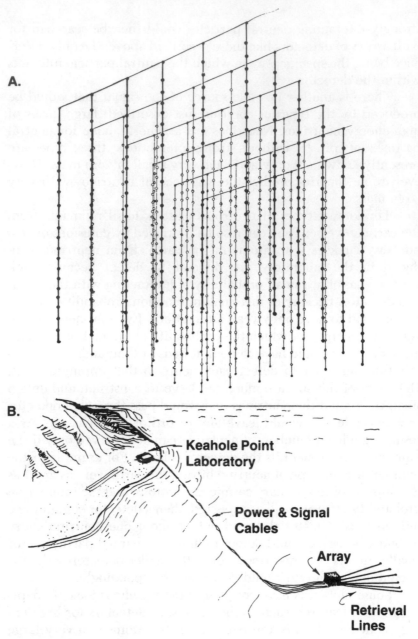

A.

B.

Keahole Point Laboratory

Power & Signal Cables

Array

Retrieval Lines

FIGURE 19   Dumand Detector of Neutrinos from Space. *A* shows a schematic representation of an array of photodetectors used to observe the effects of underwater interaction of high energy neutrinos. *B* shows how this array would be deployed in the Pacific Ocean.

ocean would sometimes interact with the atoms in the ocean and produce charged particles. These high-energy charged particles could then be detected because they would produce electromagnetic radiation while passing through the ocean. The radiation is to be observed by suspending a large number of radiation detectors, called photomultipliers, within the ocean. It has been estimated that DUMAND would be able to detect about ten interacting neutrinos hitting its volume of the ocean in a single year. Yet that small number would mean DUMAND was sensitive to a variety of hypothetical sources, such as rapidly evaporating black holes that would produce high-energy low-mass neutrinos of known types.

A potentially more interesting application of such detectors would be the discovery of fluxes of high-mass fossil neutrinos. A flux of 10^{-10} neutrinos per square centimeter per second would lead to about ten observed interactions each year, assuming that the interaction strength of these neutrinos with ordinary matter is about the same as it is for the known neutrinos. This may even be an effective way of detecting very massive fossil neutrinos, even though they are rare, provided that the total energy density that they embody is similar in magnitude to that of protons.

Another problem exists for the detection of fossil neutral particles of very low energy. For some very persuasive theoretical reasons, scientists are convinced that fossil particles of this type exist, including the three known types of neutrinos. They would remain from the period about one second after the Big Bang, when they were produced in large numbers. We have seen that if such particles exist in great numbers, then if they have even a small individual rest energy, their total energy density would outweigh that of the visible matter in the universe. This possibility has been suggested on the bases of various subatomic particle theories. It is of the utmost scientific importance to know what the main constituents of the universe are in terms of energy content. We have seen that this possibility is also relevant to the question of whether the universe is finite or infinite. Instead of allowing this question to remain in the hands of theoretical physics and theoretical cosmology, it is desirable to devise experimental methods for the direct observation of such particles.

Detecting such particles presents immense problems. Even though they may exist in very large numbers in comparison with the high-energy particles, their kinetic energies would be too small to produce ionization or observable interactions with nuclei, even if they are accelerated by galactic or intergalactic gravitational fields. The same problem exists for gravitons, the particles that go with gravity waves. Gravitons interact even more weakly with individual atoms than do most of the fossil particles that we might hope to detect. We have seen that gravity waves are to be detected through their interaction with macroscopic objects, rather than with individual atoms. Perhaps a similar approach can be taken to detect the weakly interacting neutrinos.

We can try to devise objects composed of many atoms, which will have a greater probability for interaction with these particles. Suppose that a wave hits an object containing a hundred atoms. If the object is much larger than the wavelength of the wave, then the object's individual atoms will scatter the wave independently of each other, and the total probability that the wave will scatter through a specific angle, say 10 degrees, will be a hundred times greater than the probability a single atom would give. This is called incoherent scattering. However, if a wave with a greater wavelength—one comparable to the size of the object—is scattered, then the probability of scattering becomes much greater, equal to ten thousand times that of a single atom. Such an enhanced interaction, in which the probability that the wave is scattered through some specific angle is proportional to the square of the number of atoms in the target object, is known as coherent scattering.

One of the results of quantum theory is that every particle is associated with a wave, whose wavelength is determined by the linear momentum (mass times velocity) of the particle. (The lower the linear momentum, the longer the wavelength.) Using this relation, we can determine the wavelength of any particle whose properties we know. Some of the low-mass fossil particles that would be interesting targets for a hunt could have wavelengths of many micrometers. This means that they could scatter coherently from objects whose size is many micrometers. Such objects would contain trillions of atoms. The probability of coherent scattering would be trillions of times greater than that of incoherent scattering. This enhancement in scattering could be enough to make the particles detectable. Suggestions about doing such experiments have already been made, and are likely to be carried out in the not too distant future.

The outcome of all of these fossil searches cannot be accurately predicted, but I think that we will find more than we expect—just as, in the past, we have stumbled again and again upon new phenomena when using new detecting methods. Whatever we discover is sure to extend our present theories, perhaps even replace them. I think that the fossil hunt will lead us into previously unexplored realms.

Snapshots of the Nanoworld

Human vision is one of the most effective tools for probing phenomena on a spatial scale of a millimeter or more, over time periods of about a second. Because of the structure of our eyes and the properties of visible light, vision is not effective for smaller objects and for things that change much more quickly. Also, vision is limited to the surfaces of objects that are opaque to visible light. We have been able to extend the scope of unaided vision greatly by using optical instruments (such as microscopes) and by using other forms of radiation (such as X-rays). However, many phenomena still remain hidden.

Among them are some biological phenomena happening in living cells. For example, it would be desirable to study cellular events on a spatial scale of nanometers (10^{-9} meters) and on a time scale of microseconds when division or some other crucial incident is taking place. We are currently unable to do this. In order to resolve objects of a certain size, it is necessary to use probes whose wavelengths are the size of the object or smaller. For objects of nanometer size, these probes include X-rays and high-energy electrons. Methods such as X-ray diffraction can give information about structures on a small scale, but require long exposure and work best for regular structures. Electron microscopy can also show small details. However, the objects being examined must be prepared in a way that often destroys some of what we would like to examine. The need for advance preparation also makes it difficult to follow what is happening in the cell over a period of time.

New techniques will soon overcome the limitations of spatial scale, time scale, and opacity. These techniques involve the use of intense beams of short-wavelength radiation in which a property known as the phase varies in a regular way from one point to another within the beam. Such radiation is called coherent radiation. It can be produced when many atoms emit radiation in a mutually dependent way, as in a laser. However, the problems of small spatial scale and opacity are likely to be solved only by going to even shorter wavelengths than those produced by familiar lasers. We may need to develop lasers operating at X-ray or even gamma-ray wavelengths, or possibly we will use short-wavelength radiation from other intense sources, such as a particle accelerator known as a synchrotron. What is important is not the specific way in which the radiation is produced, but rather the new types of observation that such radiation will make possible.

Imagine that we can produce X-ray beams with wavelengths of one to ten nanometers, and a duration in time of a picosecond (10^{-12} seconds). Every different type of atom in a cell or other biological system will scatter the X-rays differently, just as the different organs in the body do when photographed with conventional X-rays. As a result, the image of the cell will have a good deal of contrast; structures made of protein, for example, will be distinguishable from those made of nucleic acid or of carbohydrates. For some wavelengths, the beam would penetrate the entire cell thickness, and every atom along its path will influence the beam that emerges eventually. This means that when we detect the beam after it passes through a cell, it would contain information about what happens throughout the cell, just as an X-ray of the human body can show all of the organs along its path, not just the ones near the surface. While the cell itself is unlikely to survive long exposure to the X-ray beam, no damaging advance preparation procedures need be done, so it can be examined as it was in life.

The short duration of the beam will allow us to get information about what happens inside the object at very precisely defined times. By pulsing the beam, we can detect events separated by short intervals. For example, it might be possible to see the sequence of steps by which a protein molecule folds itself into its biologically active form after it is synthesized. The small wavelength of X-rays makes it possible to observe parts of the object that are no greater than atomic in size, which should include all of the objects of biological interest.

An especially interesting technique that might be used with short-wave length coherent radiation is holography, which was actually proposed for use with electron beams long before it was applied to the visible light produced by lasers. In holography, a beam of coherent radiation hits an object and both the intensity and the phase of the scattered beam is recorded. Ordinary recording techniques, such as exposing a photographic plate to a noncoherent beam, store only the intensity of the beam, so that its phase is lost. As a result, much of the information about the three-dimensional structure of the object is not contained in an ordinary photograph.

One way to store information about phase as well as about intensity of the scattered beam is to allow the scattered beam that emerges from the object to focus at the same place as a reference beam of coherent radiation. The two beams will produce an interference pattern where they overlap. The intensity of this interference pattern can then be recorded, giving what is called a hologram, and the pattern will include information about the phase of the scattered beam [Figure 20]. From this information, it is possible to reconstruct the full three-dimensional structure of the scattering object, either digitally through computer reconstruction, or more graphically by illuminating the recorded pattern with coherent light and viewing the resulting image. If the coherent light used to produce the image is visible light, which has a much longer wavelength than that used to record the pattern being viewed, then the image produced can be quite large. One might be able to magnify a holographic image of a bacterium up to the size of a Great Dane, with no distortion and with full three-dimensional information.

Such holographic images might reveal new, biologically important structures or substructures on a scale smaller than those we have been able to resolve thus far. Neither theory nor observation in contemporary molecular biology rules out the existence of such minute structures. Such images could also make it possible for us to directly see aspects of biological structure and function. Many of these details cannot be obtained from existing physical methods for viewing macromolecules; we have relied instead on chemical methods. But there is no reason why biophysical methods for observing nanostructures cannot replace biochemical ones, once imaging methods with adequate resolution exist. For example, the different bases contained in nucleic acids should give different X-ray scattering, just as objects that we recognize as visually distinct scatter visible light differently. An X-ray hologram could determine the base sequence in the cellular DNA or the amino acid sequence in various proteins. We could observe them as clearly as the sequence of charms along a bracelet.

We cannot hope to understand the processes of life in detail until we know not only what is in the cell but also how these mi-

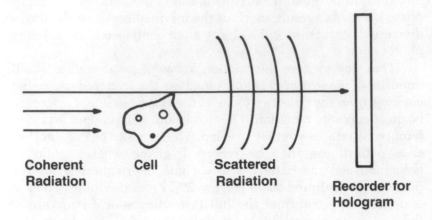

Coherent
Radiation   Cell   Scattered
                                 Radiation   Recorder for
                                                    Hologram

A. Recording a Hologram

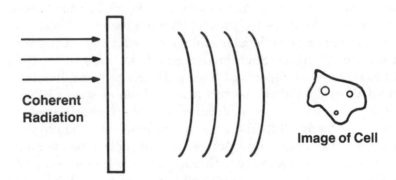

Coherent
Radiation

Image of Cell

B. Reconstructing a Hologram

FIGURE 20   Holography. A hologram is made by illuminating an object with coherent radiation, as from a laser, and recording both the incident beam and the scattered light. The recorded hologram can be viewed by passing coherent radiation through it, as in *B*, to form a flipped image.

croscopic contents change as the life processes unfold. This type of temporal structure in biological processes is probably as important in determining the behavior of living things as is the spatial structure of macromolecules.

Just as there are many spatial scales in life processes, from nanometers up to the size of each organism, so are there many temporal scales, from picoseconds up to the lifetime of the organism. We know very little about biological processes that take place very rapidly, especially when they occur over a small region of space. X-ray holography promises to provide us with this information. Unfortunately, though, we cannot hope to make holographic "movies" of living cells. X-ray holography will quickly destroy the cells on which it is carried out, so it will be impossible to observe a single cell continuously for many seconds. However, we could make a kind of pseudomovie by applying X-ray holography to a series of similar cells and combining the results to give a continuous picture.

There are two problems that remain to be solved before X-ray holography can be carried out. One is to develop a means for producing coherent beams of X-rays. A second is to find a convenient method for recording the information contained in the X-ray beam after it has traversed the target. With X-rays, the information varies over a spatial scale that is atomic in size, so small variations in the composition of the recording device can lead to a substantial loss of information. Probably the most convenient way to avoid this problem would be to use an imaging technique that spreads the scattered beam, so that points very near in the object are much further apart in the image. Alternatively, we may be able to manufacture substances with a prescribed atomic structure, optimized for recording such information. Altogether, it does not appear that there are insolvable problems in developing X-ray holography. We can expect that it will soon become a powerful technique for revealing structure and behavior on smaller scales in space and time than have been possible until now.

The new observational and measurement techniques that I have discussed in this chapter are just a few of many that scientists are currently developing. As we have seen, this relation between new techniques and new discoveries is a very old one. A much newer process—one that may ultimately influence the pattern of new discoveries as much as new techniques do—involves methods for the storage and analysis of the data obtained from experiment. These new approaches to processing data are closely connected with the use of computers in scientific research. This topic is explored in Chapter 7.

Part 3
A Fertile
Borrowing:

Outside Influences on Science

Mathematics: The Classical Language of Space, Time, and Change

New observational and experimental techniques change science from within, but there are other intellectual forces that act to change science from the outside. One of these, mathematics, is as old as science itself. Another, in use for less than fifty years, is the electronic computer. I consider these to be "outside" forces because in each case, there is a large group of people working to advance the discipline who are not primarily concerned with its possible relation to science. In many cases scientists simply adapt these outside developments to their own work. They are, in effect, consumers of mathematical methods and computer technology.

This does not mean that individual scientists are not free to develop their own branches of mathematics or their own computer systems. It is just that as predeveloped mathematics and computer systems become more common, scientists find it more convenient to use them than to develop their own. Of course, in order for this to happen, new developments have to become readily known and available to scientists. For mathematics, this has not always been the case. However, the level of mathematical sophistication among scientists appears to be rising; more scientists have studied aspects of mathematics that have no immediate application to science. This increase alone will lead to a greater use of mathematics in future science.

It remains to be seen how novel computer systems will get into the hands of scientists. In this case, the profit motive will probably be an important factor. Even though scientific applications are not the only source of potential profit, they often pave the way for other uses, and computer designers will therefore find ways to keep scientists abreast of new developments. It is notable that a large number of people originally trained as scientists are making their careers as computer systems designers. A continuing flow of new computer systems of use to science should result.

There is a tendency to think of computers as devices for doing mathematics, but this is only partly true. Therefore, though I will point out areas of overlap, I will discuss these two external forces on science independently.

It cannot be said that mathematics has always aided the progress of science. If the basic laws of some area of science are not well understood, then attempting to give a mathematical formulation to the discipline is often a sterile exercise. Nevertheless, there is some truth to the notion that the more mathematical a discipline becomes, the more nearly it approaches the character of a science. Today mathematics is used not only in astronomy and physics but also in chemistry, biology, and even the social sciences.

If the basic ideas of a science can be expressed in mathematical terms, it becomes easier to test their implications and predictions. In the physical sciences, the use of math has become essentially automatic; the fundamental laws are almost always expressed in some mathematical form at the outset. When physicists discuss concepts with one another, the discussion is almost always presented in a mathematical form.

Biologists rely less on mathematics to formulate and test their hypotheses. One reason is that, at present, physical scientists know more of the fundamental laws of their subject matter than do biologists and are more able to "control" their subject matter. Physical scientists can, in many instances, produce small variations in some system, and then use their theories to predict how the system will behave. For example, chemists can try to predict how the rate of some reaction will change as the concentration of the reacting chemicals is varied. Biologists have not had the capacity to do this as readily, but they are now developing it—for microorganisms—through some of the techniques of molecular biology. I expect that

160

biologists will be making more systematic use of mathematics for such prediction in the near future.

Nevertheless, the use of mathematics for experimental prediction should not be overrated. If a great deal of oversimplification is necessary in order to formulate ideas mathematically, then these ideas are likely to fail the test of experimental confrontation. This is especially true in the social sciences, where we do not know the basic laws well enough to make a worthwhile mathematical formulation.

Mathematics is not only a tool. More importantly, it is a language. Through this language, scientists can recognize relationships between ideas that might otherwise be hard to uncover. For example, the laws of motion in Newtonian physics were expressed originally in a form whose application required an exquisite understanding of Euclidean geometry, available perhaps to no one other than Newton himself. In the following century, when the laws were expressed in a more convenient mathematical form, using the calculus that Newton helped invent, it was readily apparent that these laws also contained the conservation laws of energy and momentum, ideas with implications beyond Newtonian physics.

That mathematics is a language is especially relevant to the creation of new scientific theories. A scientific theory usually grows in stages. Scientists must create the theory from what is partly known and partly unknown. A mathematical formalization of the part of the theory that is already known gives scientists a coherent way of thinking, and it suggests areas in which the theory is incomplete and where modifications will therefore be necessary.

In the nineteenth century, when Maxwell was devising his unified theory of electromagnetism, he began with an existing set of laws that described what was already known about the forces of electricity and magnetism. Maxwell was able to show from a specific mathematical formulation of these laws that one of them needed modification, because when it was combined with the other laws it contradicted the well-accepted principle of conservation of electric charge. The modified law appeared strange to many scientists of the time, such as Kelvin, but it was later found to be correct.

If similar triumphs are to occur in the biological sciences, biologists will have to dispense with the idea that purely theoretical analyses, of the type common in physical sciences, are less worthwhile than experimental investigations. It is unlikely that anyone coming from outside the field of biology could persuade most biologists that some specific mathematical contribution that he or she has made is of value to them. (Even in physics, a high regard for math does little to avert a general mistrust of mathematicians who try to say something about physics!) It will probably, therefore, be biologists trained in mathematics, rather than mathematicians trained in biology, who will successfully apply mathematics to biology. More young biologists are receiving advanced mathematical training, and this will eventually produce biologists who are sufficiently familiar with mathematics to be able to apply novel forms of it to the understanding of biological phenomena.

Before the late nineteenth century, developments in mathematics were often spurred by the scientists' need for a suitable language to express their ideas. In this century, mathematics has taken on more of a life of its own, and in many areas has progressed independent of science. As a result there are now a number of novel mathematical structures available that could possibly apply to new areas of science, even though many of these structures were developed with no thought of such applications.

Most nonmathematicians think of mathematics as largely concerned with numbers. However, many of the newer structures invented by mathematicians do not involve numbers in any direct way. Instead, they involve such questions as how objects can be connected to each other and what kinds of distortions can be applied to a space without changing its essential properties (studied in topology), and what rules for combining abstract symbols lead to structures with interesting properties (studied in algebra and logic). I believe that it is largely from such nonnumerical branches of mathematics that future applications will emerge, especially in the nonphysical sciences.

In sciences such as biology, it is rarely the precise numerical value of some properties of a system that we wish to understand, but rather some gross features of the system, such as the pattern of markings on a peacock's feathers or the sequence of steps that leads from a fertilized egg to a baby. It seems plausible that a nonnumerical mathematical structure will turn out to be the proper way to describe these features of biological systems. This would be a much more profound application of mathematics to biology than any done up until now.

In biology, chemistry, and some parts of physics there are

phenomena much too complex to be understood by directly analyzing them in terms of the subatomic particles of which they are composed. The laws of atomic physics are useless in these cases—but that does not mean that we cannot hope to understand the phenomena at all. A way to approach this problem is to follow a strategy, in which we formulate laws describing a phenomenon in terms of a model describing how the process under investigation occurs. Such models need not be deducible from the fundamental laws of physics, although they must be constrained by the implications of physical laws. It is in this process of model building that I believe that mathematics will play its most important role in the future of science.

Every branch of mathematics has certain structures that are special to it. The branch known as calculus deals with functions, which can be thought of as related pairs of numbers in which a change of one number causes a change in the other. Geometry deals with lines, planes, and other figures. When, within an area of science we can identify some element with a mathematical structure, the results already known to the mathematician can be used to understand the behavior of that element. For example, Newton was able to identify the motion of a body with a mathematical function, which depended on how much time had elapsed during the motion. Newton could then use what he and others had learned about such functions to describe motion. We have seen that more recently, physicists have discovered that the symmetries that apply to subatomic particles can be considered as the elements of what mathematicians call a group. When this identification was made, it became possible to use some of the ideas of group theory to find relations among the properties of these particles. That these new relations are found to be true then confirms the idea that a description in terms of groups is a natural one for these particles. I believe that such natural descriptions are likely to be discovered in other branches of science as well, and they will allow for an understanding of phenomena that now seem impossibly complex.

This searching for mathematical structures does not always work. In the sixteenth century, Kepler suggested that the orbits of the inner six planets could be identified with certain geometric constructions involving the five regular solids of Euclidean geometry. This suggestion did not survive more accurate data on planetary orbits (which Kepler himself helped provide). A more recent example, which may be more successful, was the recognition by mathematicians that there are only three types of space whose curvature is the same everywhere [Figure 21]. In twentieth-century cosmology, this idea has been applied to the overall structure of our universe, and observation should eventually tell us which, if any, of these spaces we inhabit.

Describing Change

Recently there has been a most interesting attempt to find a natural mathematical description that would apply to phenomena in diverse fields of science. René Thom is a prominent French mathematician who has made important contributions to a branch of mathematics known as catastrophe theory. Catastrophe theory studies the mathematical properties of certain modes of discontinuous change ("catastrophes"). For example, the boiling of water, which occurs abruptly when the water temperature reaches 100° Celsius at sea level, exhibits this discontinuous change [Figure 22].

Thom has proposed that some of the types of discontinuous change that are studied in catastrophe theory can be identified with some aspects of morphogenesis in the embryonic development of multicellular organisms. We have seen in Chapter 4 that the understanding of development is one of the outstanding problems in biology. One essential ingredient needed to solve this problem is the right picture for development; without it, the many facts that experiments reveal about development will not form a coherent pattern. It is possible that the right picture will someday be expressed in traditional scientific language, such as the physical picture used in molecular biology. But I think it more likely that a creation of some mathematician will provide a key to unlock the secrets of development, and that this will be one of the great triumphs of science.

Thom and his colleagues are following up on ideas first stated early in the twentieth century by the British biologist D'Arcy Thompson. Instead of focusing on the biochemical mechanisms through which development takes place, they are trying to understand morphogenesis by using results that follow from the fact that living things develop in a three-dimensional space with specific geometrical and topological properties. Thom is not saying that biochemistry does not constrain how morphogenesis occurs, but rather that it is possible to understand some aspects of morpho-

Spherical Surface

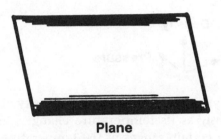

Plane

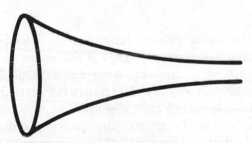

Pseudospherical Surface

FIGURE 21 Spaces of Constant Curvature. There are three types of mathematical space whose curvature is the same at all points. In two dimensions these are the surface of a sphere, whose curvature is positive, the plane, whose curvature is zero, and the pseudosphere, whose curvature is negative.

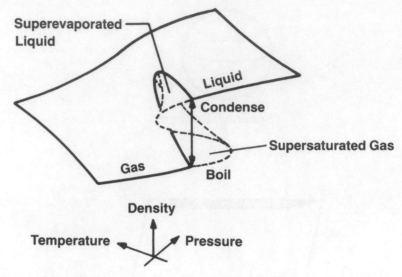

FIGURE 22   One Type of Catastrophe. The density of a substance can vary continuously as its temperature changes, following the smooth surface. For certain ranges of the temperature, two values of the density can exist and the substance can pass abruptly from one to the other (in the phenomena of boiling or condensation), as indicated by the double-pointed arrow.

genesis, and to find a natural language to describe it, without knowing the biological details. This is analogous to the fact that one can understand a computer program without understanding the solid state physics and the electronics that actually govern the working of the computer's components.

The attempts to apply catastrophe theory to embryonic development is only part of a very ambitious program. Thom is trying to construct a complete description of forms and how they change in time. By "form," Thom means any object that seems stable to human perception, especially those that exist in the macroscopic world. Examples of forms are tables, living organisms, and planets. Forms include most of the objects of concern to science, except perhaps the subatomic particles. In this general program, as in the specific application to development, Thom is basing his understanding of the properties of forms on the fact that the forms exist in a four-dimensional space-time. One of the main results of catastrophe theory—that only a small number of distinct types of discontinuous change can exist—is a consequence of the geometrical properties of *three*-dimensional space. If space had a different number of dimensions, there would be different types of change allowed.

Simpler examples of the consequences of the three-dimensional geometry of space have been known for a long time. It follows just from the geometry of space that of all objects with the same volume, a sphere has the smallest surface area. This is the reason that many objects in nature have spherical shapes, whatever they are made of, because the physical processes that govern their shape work to minimize their surface area.

What Thom is proposing is analogous to the branch of physics known as thermodynamics, which describes aspects of the behavior of bulk matter that do not depend on the composition of the matter. For example, many gases cool in the same way when they expand. Thom has used some of the discoveries in modern mathematics to describe a wide variety of other phenomena involving forms and how they change. He believes that these phenomena can be understood qualitatively, on the basis of the general principles of catastrophe theory. One of the things that he tries to explain is why forms exist at all, which he calls the problem of structural stability. It is too early to tell how successful his program will be. Many mathematicians and scientists object to his claims. However, I am convinced that whatever the fate of catastrophe theory, Thom's proposition that some of the behavior of objects is a consequence of universal mathematical relations is likely to survive and to provide important insights for future scientists.

There are other places in science where such ideas will probably bear fruit. One of the important questions about proteins is how they arrange themselves in three dimensions. These arrangements are crucial to the functioning of proteins as catalysts for chemical reactions, as well as for the antibody functions that play important roles in the immune system. It seems likely that the arrangement of proteins is highly constrained by general principles about how things can be connected in three-dimensional space, and that many of the details of this arrangement do not depend on the biochemistry of the protein. If so, then a mathematical study of the possible ways in which long polymers can arrange themselves into three-dimensional arrays should cast much light on how proteins function, on why proteins are the preferred molecules for catalytic activity in living things on earth, and on whether alternative molecules or other structures might perform similar catalytic functions under different conditions.

The Origin of Complexity

Most of the matter familiar to us consists of atomic and molecular combinations of neutrons, protons, and electrons. The complex internal structures of neutrons and protons are irrelevant to most of the properties of matter that we are aware of, just as the internal structure of bricks is of little importance in determining how effective bricks are for making a house. Furthermore, under the conditions found in ordinary matter, the basic equations of quantum theory, which describe how the constituents of matter behave, are relatively simple. There is little reason to think that under ordinary conditions, such simple equations will require modification in the future.

Nevertheless, in spite of the simple structure and laws that govern the constituents of matter, large-scale objects often display complex behavior whose explanation is not intuitively apparent. This is especially true in systems that contain many atoms or other components that strongly influence one another. Examples of such systems are fluids in motion, multicellular organisms, the photons in a laser cavity, and the neurons in a brain.

It is difficult for the human mind to classify and understand all the different types of complexity that occur in large systems. Two types of such complexity, however, offer some prospect for understanding through develements in applied mathematics and computer analysis. These two types of complexity are usually thought of as opposites and go under the names "chaos" and "order." Yet, there are indications that certain kinds of chaos and order are consequences of the same type of mathematical structure, known as nonlinear equations. Perhaps this is a modern version of the ancient philosophical view of the Greek philosopher Anaxagoras, who believed that opposites emerge from a formless Absolute, and eventually return to it.

Chaos

Even in large-scale systems (in which quantum theory is thought to play little role) behavior is often unpredictable and subject to large variations as the result of what appear to be minor changes in the system. We refer to this as "chaotic behavior." For

168

example, a liquid may flow smoothly and predictably for a certain range of velocities but becomes turbulent when the velocity increases slightly. This kind of random behavior mocks the words of the nineteenth-century French mathematician Pierre Laplace, who suggested that if we knew the present well enough we could predict the future exactly.

Chaotic behavior *cannot* be predicted, even when the fundamental equations are deterministic, as are those of Newton's physics. Maxwell already pointed out in the mid-nineteenth century that some phenomena that are governed by deterministic equations are nevertheless unpredictable in practice because the way the phenomena develop depends critically on the precise situation at the start of the development. Because we can only make measurements with imperfect accuracy, we never have a perfect knowledge about the conditions that describe any system. For systems that behave chaotically, the smallest error in our knowledge of their condition when we begin to study them will lead to complete uncertainty about conditions after a short time.

Chaos is not universal in everyday phenomena, therefore it is important to understand when and how it arises, and to understand precisely what is involved. It has been found that chaos can be studied in model situations when the equations are simple enough to examine in great detail. This approach may soon answer our questions about chaos. We may not be able to exactly predict next week's weather but we will know why we cannot do so.

The models of physical systems that are helping us to understand chaos involve nonlinear equations, which describe how one quantity or several interacting quantities change with time. The main characteristic of such nonlinear equations is that the quantities in the equations vary with time in a way that depends acutely on their amount. For a nonlinear equation if we double the amount of the quantity that is originally present, the amount present at a later time will not simply double, but will change in some complicated way. The nonlinear equations themselves are deterministic, which means that if we know the amount of the quantity present at the outset *precisely*, then the amount present at any later time will also be fixed precisely by the equations. This was the standard behavior of physical systems in pre-quantum physics, and was Laplace's inspiration.

What has emerged from the recent studies is that a system whose change is governed by a nonlinear equation can undergo several qualitatively distinct behaviors, depending on the numerical values of certain numbers, called parameters, that appear in the equations. One such number could, for example, represent the extent to which the equation is nonlinear. For some values of the parameter, the evolution of the system is periodic, like a planet orbiting the sun, and is therefore predictable. However, a change in the parameters that define the equation leads the system to display a completely different behavior. Instead of following a periodic orbit, the object would follow a complicated path through its possible situations, perhaps taking on every imaginable value, without appearing to return to its original configuration. Which of the many such possible paths the system takes depends critically on its initial configuration. A minor change in this configuration can lead to its following completely different paths. This is chaotic behavior.

Sometimes, the system eventually returns to periodic behavior after a long spell of chaos. There are other cases, also based on deterministic equations, in which true randomness may occur in the motion, so that it never returns to being periodic. Instead, the motion may eventually confine itself to a region of space whose dimensionality is between that of a surface and a volume. Such peculiar dimensionalities, which the mathematician B. Mandelbrot has called fractals, seem to pertain to a number of scientific phenomena, such as phase transitions. Fractals may be an important subject in the applied mathematics of the future.

It is irresistible to speculate that chaos emerging from order in this way is the rule in all cases where randomness occurs. This is probably true in those situations governed by pre-quantum physics, such as the turbulent flow of a liquid. Chaos may also be relevant to the problem—the direction of time—discussed in Chapter 3, about why order tends to decrease spontaneously. Systems whose order has decreased may be those whose atoms are undergoing chaotic motions that cause the microscopic order originally present to rapidly disappear. Questions such as this will form an important area of study for mathematical physicists in the future.

It is tempting to think that even the randomness inherent in quantum theory is actually a chaotic manifestation of underlying determinism. There have been some ideas along these lines, but none have led to any new suggestions of observable phenomena. Furthermore, even if the randomness of quantum theory does turn out to be the type of deterministic chaos that occurs as the result of some nonlinear equation, it would not allow us to make predictions about atomic phenomena that are any more accurate than those allowed by the true randomness that is built into the equations. Nevertheless, such a discovery would be very important, as it would considerably change our view of how randomness enters into nature.

Order

Ironically, order can come about from the destruction of an underlying simplicity in which no order is apparent. The problem now under discussion, of how order develops in physical systems, differs from the one in our discussion of the direction of time. For an isolated system, it is empirically true that order cannot increase (the law of entropy increase), but the systems we will discuss now are in contact with an environment with which they can exchange matter and energy. Under these circumstances, nothing prohibits an increase in order, and in many cases that is what happens. What needs to be understood are the mechanisms by which order can arise under these circumstances, and the general features of such order.

While order is in some sense the opposite of chaos, there are nevertheless some similarities in the ways that both occur in physical systems. In each case, what often happens is that there is a sudden change from one situation to another. On one side of the change, there is a radically different behavior than on the other side. These are like the phase changes discussed in Chapter 1. For example, when a laser begins to operate, the light in the laser cavity changes abruptly from a mixture of many different wavelengths to a single dominant wavelength. This behavior is analogous to what happens when a piece of iron becomes magnetized. In that process, there is a rapid change in the direction of the large number of atomic magnets that compose the piece of iron. Before the magnetization, these directions are arranged haphazardly; afterward they are almost all the same. It is usually through such a change in a system's bulk properties, rather than in the arrangement of its microscopic constituents, that the onset of order is recognized.

It would help our understanding if we could find general rules that describe the onset of order in many different physical processes, such as the operation of a laser and the magnetization of iron just described. Even though the specific processes that are taking place are very different, it may be that the mathematical equations that describe these processes are similar. We know of cases in which similar equations describe phenomena in which the constituents are different. Sound waves in the air and light waves in a vacuum satisfy similar equations, even though in one case the equations describe moving atoms, and in the other electromagnetic fields.

One thing that has been discovered is that various changes from disorder to order can be described as consequences of a particular type of mathematical structure, called nonlinear differential equations. These are similar to the equations discussed in connection with chaotic behavior. Again, the main characteristic of such equations is that the way in which the quantities described by the equations vary in time is very sensitive to how much of each quantity is present. This is different from the linear differential equations that describe phenomena like sound waves, in which the rate of change of various quantities is less dependent on what is already there.

Some of the solutions to these nonlinear equations describe situations in which order arises. The form that a solution takes can depend critically on the value of certain numerical parameters that appear in the equations. For example, one parameter in the equations describing magnetization is the temperature of the magnet. It often happens that when the parameter lies in one numerical range, say corresponding to high temperatures, the individual solutions to the equations are symmetric. They describe a situation in which there are equal numbers of atomic magnets pointing in any direction in space. For a different range of values of the parameter, corresponding to low temperatures, the individual solutions may not possess this property, and some directions for the atomic magnets are favored over others. This can happen even though the equations themselves do not favor one direction over another.

The change in the form of the solutions often signals the onset of order in the system. For a ferromagnet, there is no overall magnetization when the temperature is high, because the magnetic effects of the individual atoms cancel one another. At low temperatures, these atomic magnets are lined up, and the whole magnet becomes magnetized. The physical phenomena occurring inside or near the magnet now depend on direction in a way that was not the case at high temperature. For example, if a compass needle is brought near the aligned magnet, the needle will swing toward it, which would not happen when the individual atomic magnets pointed randomly.

The situation in which solutions to a set of equations have less symmetry than the equations themselves is similar to the situation for quantum fields, discussed in Chapter 1. The fields are also described by a type of nonlinear equation, and broken symmetry occurs when a parameter in the equation lies in a certain numerical range.

Another qualitative feature sometimes appears when several nonlinear equations are considered together. Such a system of equations is relevant when different quantities are needed to describe the system—for example when there are concentrations of several different interacting molecules in a solution, or when there are photons of various wavelengths, as in a laser cavity. The laser light may begin as a mixture of wavelengths, but the competition described by the equation may quickly result in the rapid growth of only one wavelength, while the others die out. This is reminiscent of how some biologists interpret the Darwinian idea of evolution through natural selection. Precisely which solution of the equation wins out in the competition depends both on the initial conditions and on the environment. But one solution will dominate in a short time, over a wide range of conditions; that is a property of the equations themselves, a property that is shared by the nonlinear equations that describe a variety of distinct phenomena.

The mathematical study of complex behavior is just beginning, but already it has illuminated a number of dark areas in science. Perhaps the most sweeping result will be a general classification of the types of complexity, such as chaos, together with a demonstration that a relatively small number of these types, when suitably combined, can account for a wide variety of behavior in large-scale systems. What I have in mind is analogous to what happened in the sixteenth and seventeenth centuries, when motion was first analyzed correctly by Galileo, Newton, and others. Before that, it was not clear how many types of motion there were in nature or what relations existed among them. Through these analyses, it was recognized that calculus, then a newly invented branch of mathematics, held the key. With calculus it becomes possible to determine all forms of motion; by appropriate combinations of a few of the simpler forms of motion, most of the types of motion observed in nature can be described. In the case of complexity, we might learn something analogous through mathematical studies of the types of equations that lead to order and chaos. If a general description of complexity does arise from such approaches, this would be an important new example of how developments in mathematics can lead to advances in science.

New Descriptions of Space-Time

On occasion a specific mathematical structure reveals a totally new aspect of a natural phenomenon. This happened with the discovery of non-Euclidean geometry as an independent mathematical structure in the nineteenth century; Einstein later realized that this geometry applied to our own space-time. Perhaps the most striking aspect of this discovery is the possibility that our universe is finite and resembles the surface of a sphere rather than a Euclidean plane, but in one more dimension.

Emboldened by Einstein's success, many mathematicians and physicists in the 1920s and '30s tried to use other mathematical structures to describe nature, but with no significant results. Because of this failure, the use of abstract mathematical structures in physics became a marginal pursuit for many years. This situation has changed recently, both as a result of the success of group theory in particle physics and because of the efforts of prominent scientists working in the field of general relativity to expand the use of novel mathematics. It may be that the mathematics that we use to describe objects in space and time, which comes from our experience with everyday objects, is not the most suitable to use for subatomic particles or for the cosmos. The use of other mathematical languages could help answer some questions in these fields.

One step in this direction has been taken by the British mathematical astrophysicist, Roger Penrose, and his collaborators. They hope to find a replacement for the conventional space-time description used in all physical theories, in which each event occurs at a point that is described by four real numbers—the three space coordinates and one time coordinate of the point. Penrose and his colleagues label and describe where and how events happen by the use of a new type of mathematical structure: the twistor. A twistor consists of two sets of complex numbers. Each complex number consists of a real number and an imaginary number. There is a complicated relationship between the complex numbers in the twistor corresponding to an event and the space-time coordinates of that event. Actually, twistors correspond more nearly to sets of points whose coordinates are complex numbers rather than real numbers. The set of points can be thought of as falling along the trajectory of a particle that moves at the speed of light through space-time. Instead of thinking of space-time as made up of an infinite number of points, as in the usual description, the twistor description pictures space-time as constructed from an infinite number of criss-crossing trajectories.

Roger Penrose has many ideas about what will emerge from the use of the twistor description. One is that since our usual space-time description is a derived rather than a fundamental concept, some of its specific properties, such as the precise number of dimensions, could be inferred from the properties of twistors. That is, it might be that starting with twistors, we could reconstruct a four-dimensional space-time, but not a space-time of a different number of dimensions. Penrose also believes that the twistor description forms a more natural link between special relativity theory and quantum theory than does the usual quantum theory of fields. He suggests that when such a link is completed, the usual space-time description will be only an approximation and that some of the difficulties that arise in quantum field theory (such as the occurrence of infinite values for some quantities) will be avoided in the twistor description of particle interactions. These hopes remain to be realized.

If it turns out that the twistor discription is applicable to nature, then it may have other far-reaching consequences. Just as quantum fields manifest themselves in ways other than as subatomic particles, so it is likely that the twistor description contains structures that describe phenomena other than subatomic particles. For example, the twistor description might imply a more detailed connection between different regions in space-time than does the usual four-dimensional point description. This is plausible because twistors correspond to trajectories, which link points in widely separated regions. If so, this might shed light on a problem in present cosmology. According to the usual description, the properties of matter in regions that are far apart in space, but near in time, should be completely unrelated. Yet there is convincing evidence, from observation of the light from distant galaxies which lie in opposite directions in the sky, that conditions in such regions are similar. For example, quantities such as the ratio of electron mass to proton mass, which could in principle vary from place to place, are the same, to a high degree of accuracy, in regions that are very far apart. The puzzle is why this should be so. Alan Guth's inflationary cosmology gives one answer to this question, as it implies that all such regions that we can now observe were in close contact in the early universe. However, this may not be the right reason. If twistors provide a more fundamental description of events than do space-time points, it may help solve this puzzle—among others.

If twistors do give a more fundamental description of the universe, for which the space-time coordinate description is only an approximation, then there is reason to examine all of the structures that occur naturally within the twistor description, to see which of them might correspond to other, unknown aspects of nature. In order to do this, mathematical physicists will make use of some branches of mathematics that overlap with the twistor description, such as topology, but that have played little role in conventional physics so far.

This area of mathematics may play an important role in future science even if twistors turn out to be irrelevant. I mentioned in Chapter 3 that we do not know how space-time is connected over very small distances and times. This gap in our understanding is accentuated by an idea that emerges by combining quantum theory and general relativity, as has been pointed out by the American theoretical physicist John Wheeler.

In any quantum field theory, the fields undergo a process known as "vacuum fluctuations." These are just another manifestation of the uncertainty principle; here it has been applied to field strengths rather than to particle positions. Because of vacuum fluctuations, the strength of any field, such as the electric field, in any region of space changes rapidly, especially over short intervals of time. In a quantum version of the general theory of relativity, still to be worked out, fluctuations in the gravitational field would occur. This would mean that the properties of space and time themselves undergo such fluctuations. A region of space that is flat at one instant may become curved at the next instant.

These fluctuations in space-time include the way in which regions connect to one another. A vacuum fluctuation can convert a small region that is smoothly connected, such as the surface of a sphere, into two unconnected regions such as two spheres. When viewed over a sufficiently small scale of 10^{-33} centimeters, the structure of space-time is likely to be extremely complicated. It might look like a liquid that foams because its motion has produced many air bubbles. It may be that this structure arises because of vacuum fluctuations, or it may be that space-time has an intrinsically complicated structure. In either case, the way that space-time is connected over very small regions is likely to have important implications for particle physics and for the description of black holes. In both these cases, many of the unsolved problems involve what happens at small space-time separations. We have seen that it may be possible to solve these problems without assuming that the points of space-time are connected in a complicated way, if, for example, space-time is discrete. Nevertheless, it is worthwhile for scientists to look at approaches that involve novel connectivities for space-time.

A precise description of how space and time connect will require more abstract and powerful mathematical techniques than have been used in physics up to now. Mathematicians working in the field of topology have been studying how to describe abstract "spaces" with complicated connective structures. I do not know whether it will be possible to directly apply what they already know or whether some further mathematical advances will be needed. But it is very likely that the development and application of a suitable mathematical representation to describe the small-scale structure of space-time will be one of the important problems in the mathematical physics of the future.

Notions of connectivity and topology may have other important roles in future physics. The quantum theory explains the "quantization of spin," that is, the fact that spin can take on only certain values whose ratios are simple rational numbers. The explanation is based on the mathematical properties of rotations in space. Recently, explanations have been given for the fact that the electric charges of subatomic particles are quantized, using similar reasoning based on the mathematical properties of the internal symmetry group that describes fundamental particles. Several generations of physicists have come to regard such arguments as a natural explanation of quantization and of the specific numerical values that the quantized property can have. But we are just beginning to realize that these arguments also involve subtle topological assumptions, which may not always be satisfied.

It is conceivable that there are circumstances in which these additional topological assumptions do not apply, because of peculiar space-time connectivities, and where the usual conclusions about quantization of spin and charge may be untrue. We may someday discover objects with electric charge or spin that are impossible by the standard arguments. Anticipating this would be a spectacular consequence of the application of topology to physics.

Why Can Mathematics Be Applied to Science?

Why should mathematics be effective in science at all? That is, why are we able to describe natural phenomena through the use of concepts originally devised by mathematicians? The question of the effectiveness of mathematics is especially puzzling with the use of "preformed" mathematical concepts such as groups, which were not originally invented with scientific applications in mind. Sometimes scientists approach their subject matter by methods that are later found to be suspect. There was a long period in the eighteenth century when scientists regularly used theological arguments to draw conclusions about physical laws. (Ironically, they often reached correct conclusions on the basis of arguments that we would now consider irrelevant.) Although mathematics has been used with apparent success in science for centuries, we should consider the possibility that there is no real connection between the two, and that mathematics will not continue to be used in future science.

I think, however, that the notion that mathematics is really irrelevant to science, or that it will become less relevant in the future, cannot be defended successfully. It is hard to imagine science without the systematic use of symbolic representations—and the analysis of symbolic systems is the essence of mathematics. This is especially true in fields such as physics, where the aspects of nature now under study are less amenable to ordinary intuition and the use of symbolic representation is essential. But mathematical reasoning is likely to become essential to other fields as well, especially insofar as they will "borrow" ideas and applications from physics.

We need to understand why it is that mathematics works in science. One approach to this question can be connected with the Platonic view of the nature of mathematics that has been held by some mathematicians and philosophers such as Kurt Gödel. According to this view, mathematics is discovered, rather than created. The structures studied by mathematicians are not simply products of the human mind, but rather exist in an independent "universe," which mathematicians explore through their work; this is like the exploration of the physical universe by scientists. If this argument is correct, it is possible that there would be some relationship between the two "universes," so that the structures discovered by the mathematicians would indeed be relevant to those of the scientists.

I have strong reservations about the Platonic view. The process and the course of mathematical discovery seems so clearly tied to the idiosyncrasies of human thought that it is difficult to believe that mathematicians are exploring a world independent of the human mind. Furthermore, the mathematical universe, unlike the physical world, cannot be explored by both thought and sense; we can only approach it with our minds. In the absence of any indication that mathematical concepts originate outside of our heads, it is intellectually more justified to assume that the mathematical "universe" is created by our thoughts, rather than that it has an independent existence. Even if the thesis of the mathematical Platonists is correct, it would explain our ability to apply mathematical structures to science only if it is true that the number of structures that exist in the mathematical universe is relatively small and that we have already explored most of them. Otherwise, it is unlikely that the parallel exploration of two independent universes would uncover structures in one relevant to the other.

There are circumstances within some areas of mathematics in which there are a limited number of structures that might be applicable to nature, and where an examination of all of these is feasible. For example, I have pointed out that there are only three kinds of space of constant curvature, so that if our universe is a constant curvature space, it must be one of these three types. However, there is no indication that the total number of conceivable mathematical structures is very small, or that we are yet anywhere near knowing a major fraction of the possibilities. If we had already discovered most conceivable mathematical structures, the rate of mathematical innovation would be slowing down—and there is no indication that this is happening. For all these reasons, I believe that we cannot look to mathematical Platonism for an explanation of the effectiveness of mathematics in science.

An opposite approach stresses that the ideas of mathematics and science are *both* created by the human mind. If so, then we should not be surprised that two different sets of ideas should be closely related, for they both have the same origin. There is an element of truth to this view, but if taken to extremes it would imply that our scientific view of the world would ultimately be determined by what is in our heads, rather than by what is outside of us. I cannot accept this idea for science, although I believe it is true for mathematics. We have been surprised so often by what scientists discover that there must be a component to science that goes beyond the specific ways that human beings think.

This "subjectivist" point of view about the relation of science to mathematics may be most relevant when there is more than one possible theoretical analysis of the same body of scientific knowledge. It is often found that while some area of science was described originally with one type of mathematical language, it is equally possible to use one of several other languages. Sometimes, there is a shift from the original description to a new one. One example was the shift from a geometric to an analytic way of formulating Newtonian mechanics. Sometimes, an alternative mathematical description is treated as a curiosity and is not adopted by scientists until much later when it is found that it is preferable in the light of new discoveries.

My own view about the reason for the effectiveness of mathematics in science is closer to the subjectivist view, but not identical to it. It begins with the notion that the concepts of both disciplines grow out of a set of intuitions that originate in the same kinds of human experience. While there may be an immense number of logically consistent symbolic structures, mathematicians tend to be most interested in those that have shown up "naturally," in the course of the development of mathematics itself, rather than structures invented independently. Pure mathematics tends to grow somewhat like a tree, with new branches originating out of problems that have come up in existing branches. Furthermore, when mathematicians invent novel symbolic structures in order to solve such problems, they generally use concepts that are directly abstracted from everyday life (although they use them and combine them in ways that have no simple relation to how they occur in ordinary experience). For example, group theory is based on a concept of "multiplication," abstracted from the everyday experience of performing, in a definite order, two related activities, such as two distinct motions of the same object.

Much the same is true for the concepts of science. Even when scientists think about phenomena very different from those of everyday life, such as subatomic particles, they do so by combining in novel ways concepts derived from everyday life. For example, one of the true novelties of the atomic world, Heisenberg's uncertainty principle, is based on the notion that when we observe something, the act of observation may change the property being observed, just as the blood pressure of a nervous person may skyrocket when the doctor approaches to measure it.

Because the basic concepts of pure mathematics and science originate from ordinary experience, it should not be surprising that the structures studied in one discipline often turn out to be applicable in other disciplines. However, as both mathematics and the sciences develop, there is a tendency for those working in each field to base their work on concepts that have been introduced previously within that field, rather than on concepts deriving from ordinary experience. These concepts are usually already fairly abstract, and are often removed from experiences that are shared with workers in other fields. Because of this, I think it likely that as mathematics and science develop, fewer of the structures of one will pertain to the other. This process is somewhat similar to what happens during biological evolution, when, as two originally related species separate, their behavior and appearance gradually become more and more different.

Nevertheless, mathematicians will continue to create an imposing array of symbolic structures, some of which will continue to play a central role in the future of science.

Computers:
Launching Science
into a New Age

Of all the changes that have taken place in recent years in how science is done, none is more fundamental and sweeping than the use of computers. Their increasing role shows no signs of abating; it is safe to predict that their use will continue to bring about startling changes in science.

Many studies have focused on the effects of computers on society as a whole, but little attention has been paid to their specific effects within science. This is surprising, since with the possible exception of video games, there is no aspect of modern life where the use of computers is more widespread.

Computers were, in fact, first developed by scientists, who have significantly influenced their evolution. What is not well understood is how the use of computers has begun to shape what scientists do, nor how, ultimately, their use changes the kind of activity science is.

There have been two main types of computer applications to science, one in experimental work, the other in theoretical investigations. Experimental scientists have made use of the ability of computers to record and analyze very large amounts of data in a short time. This has made possible many experiments that otherwise would have been beyond our analytic ability. Furthermore, the sheer amount of data that is produced by some experiments in physics and biology is so large that even storage and access could not be accomplished were it not for the data processing capabilities of computers.

In particle physics experiments, it is now possible to scan and analyze a very large number of particle collision events in order to isolate a single event with specific characteristics. In one recent experiment, carried out at the CERN laboratory in Geneva, a million examples of particle collisions were analyzed by computer almost as they happened. The computer searched for events of a certain type, which were evidence for the production of a previously undetected particle called a W-particle [Figure 23]. Five such events were eventually identified. While such searches are in principle possible for human beings to do, we could not do it nearly as quickly—and certainly not while the experiment was in progress. Before the advent of computers effective for such analysis, particle physicists employed large numbers of people as "scanners," to make such searches. The computers have essentially replaced these people, making them among the first human casualties of the application of computers to science.

Computers have also made possible a different type of experiment, in which several large bodies of data are compared systematically to search for hidden correlations. Molecular biologists have made use of this ability to look for similar genes or control elements in the nucleic acids of different organisms.

As computers become more effective, they will be able to discern correlations amid enormous amounts of data. A problem with this at present is that it is usually necessary to preprogram the computers to look for specific patterns in the data. In other words, the scientist must know in advance what he or she hopes to find. It would be more useful to scientists if unexpected results hidden in the data could be found. For example, astronomers would be interested in knowing if the radiation coming to earth from some distant astronomical object varied in intensity with the season of the year. But unless a computer that was analyzing measurements of this intensity were programmed to look for such a correlation, it would probably not be detected.

This limitation on computer analyses will be overcome as the data-analyzing abilities of computers improve, so that a wider variety of possible correlations can be investigated. Eventually computers will be able to look for unexpected relationships among

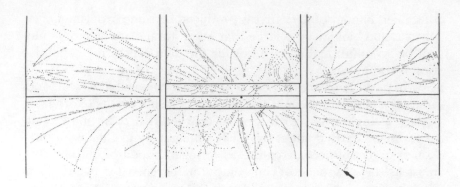

FIGURE 23   A Complex Particle Event. The photo shows many tracks of particles that are passing through a detector. The tracks have been reconstructed by computer analysis of the detector output. One of the tracks, indicated by the arrow, is the path of a decay product of a newly discovered subatomic particle.

large amounts of data. Data banks, in which large amounts of scientific information is stored in a form that is readily accessible for computer analysis, will contribute to this capability.

Computers can make it possible to carry out observations in places that would otherwise be inaccessible. Automated space probes have already made many such measurements—and some scientists argue that all of space exploration can and should be done solely with automated probes. At present, however, computers cannot control experiments where new decisions must be made over the course of the experiment. The most complete attempt to automate a critical experiment was that carried out by the Viking probe of the surface of Mars. This experiment left important questions about life on Mars unanswered, partly because there were no people present to vary some of the experiments at crucial times.

Advances in the fields of artificial intelligence, and molecular engineering could result in small computer packages with decision-making capabilities comparable to those of human beings. When this happens, the decision between manned and unmanned space probes will be made on grounds other than that of superior human effectiveness. Nevertheless, I believe that humans will not willingly surrender the exploration of space to computers, even if computers do prove to be more effective.

Simulating Nature

The computer's capacity for numerical analysis, and to a limited extent, for symbol manipulation, has been most useful to theoretical scientists. This capacity has made it possible to do calculations that involve so many different steps that it would be impossible for human beings to do them in any reasonable time. For example, the predictions of quantum field theory about the way in which electrons behave in magnetic fields have been calculated to an extremely high accuracy, one that would have been unimaginable ten years ago. In this type of calculation, no basic change is made in the physical theory, although certain approximations must be made to adapt the calculations for computers. The calculation of the behavior of an electron in a magnetic field, for example, involves a large number of complicated integrals. Using a computer, such calculations have been done by an approximation scheme known as the Monte Carlo method. The integrals are calculated by evaluating the function to be integrated at only a random selection of points within the range of variation. If the random selection of points is large enough, this procedure gives a fairly accurate value for the required integral. Such an approximation scheme is well suited for computers, which can easily perform a large number of individual numerical computations. The Monte Carlo method has now been extended to other calculations in quantum field theory, especially those in which continuous space-time is replaced by a discrete lattice. There is some indication that the properties of interacting quarks and gluons can also be understood through such calculations.

Numerical simulation methods also play important roles in experimental physics. In many experiments carried out by particle physicists, the quality of the data obtained by the detection equipment is not very precise. The data may contain large amounts of irrelevant background effects. Also, there may be technological limitations on the accuracy of the detectors themselves. Scientists need to have some idea about how their instruments will respond to the phenomena that they hope to observe. To accomplish this, they use theoretical models of the process under study to predict the type of phenomena that may occur, and then use these predictions to calculate how the detectors would respond to such phenomena. These calculations (which are also often done by the Monte Carlo method) can be compared with the actual data obtained in the experiment. This is a much more theory-bound approach to experiment than scientists are used to. Ultimately, the need for this procedure in particle physics derives from both the complexity of the phenomena and the difficulty in observing individual subatomic phenomena with instruments built on a human scale. This shift to using models has allowed scientists to extract useful data from experiments that only a few years ago would have yielded a clutter of useless information.

However, this use of theory has introduced new dangers into experimental physics. When scientists must build a great deal of theory into the analysis of experiments, it becomes less clear what has gone wrong when the actual data don't agree with the simulation. Also, there is the danger that valuable data that do not fit into the theoretical framework will be discarded because the proper methods for analyzing the data are unavailable. Nevertheless, the use of computers will continue to shift the border between theory and experiment in physics.

Computer-Based Intuition

Intuition relies on unconscious thought processes rather than on a detailed analysis of a situation. Intuition is not the result of magical inspirations; it depends on a person's experiences. When these experiences involve computer calculations, as they do for many young scientists, such calculations naturally become a source of intuition for further work, even though that work may be analytic.

A theoretical discovery recently made by the American scientist M. Feigenbaum is a case in point. With the use of a simple computer, the hand calculator, Feigenbaum discovered a mathematical phenomenon closely related to the occurrence of chaotic behavior in deterministic systems. His discovery spurred great interest in the field of deterministic chaos and this field soon became the subject of intense analytic work that has served to clarify and extend the original result.

Some computer calculations have suggested profound changes in physical theories, such as the replacement of continuous space-time by discrete space-time. For some scientists, it is a short step from introducing a model that makes some change for the purpose of computational simplicity, to thinking of the model as a fundamental part of our picture of the world.

The type of problem that is most easily solved with computers involves discontinuous change, in which both the property that is changing and the property that causes it to change jump from one value to another, without passing through the values in between. For example, this is what happens when we describe the boiling of a liquid in a model where time is discrete rather than continuous.

Because of its greater convenience for computer calculation, discontinuous change will become an increasing part of scientists' models, and consequently an increasingly important element in the intuition that many scientists have about how processes take place in nature. This intuition will in turn play its part in the development of future theories.

This evolution in our thought is not unexpected. If, as suggested by Immanuel Kant, the properties that we assign to such concepts as space are partly expressions of the way in which human beings are constructed to think, then by using new tools to help us think, we will be led to refine our description by ascribing new properties to concepts such as space. Furnishing us with new intuitions about how to think about the world may be one of the most important influences of computers on the content of science.

Reading the Genetic Code

The use of computers in biology has lagged behind physics, but this is beginning to change as molecular biologists obtain more data about large biomolecules. An important advance for molecular biologists has been the development of techniques to "sequence" nucleic acids, that is, to determine the sequence of fundamental bases along a particular strand of DNA or RNA. Anyone who systematically reads biology journals will come across articles in which the main part of the text is a long string of symbols such as ACGA [Figure 24] representing the bases in a specific section of the genetic material of some organism.

To determine such sequences, a large amount of chemical information needs to be processed. For short sequences of a few hundred units, this processing problem is fairly straightforward. But as the sequence lengths rise into the tens of thousands, as has already happened, or the millions—which corresponds to the complete genetic information of a bacterium or of a chromosome in a eucaryotic cell—the problems involved in extracting the sequence from the chemical data become more and more difficult. Computers are already playing important roles in solving these prob-

```
        10         20         30         40         50         60
TCTAGAGTGC ATCAGCTACA GTAAAGACTT GAAGGGCGAA GAAAGCTTCA TGCTGTCTAG
        70         80         90        100        110        120
CAAAAGGGGA GGGGGGCACA CGTGTGAACA CAAAGAAAAC GGAGTCTAAT ACAGGGAATC
       130        140        150        160        170        180
GAACTTTGGA CCTTCCTAAC ACAAAGCAAC CTCACTACCC AGCTGAACCA CAGAAGAAAT
       190        200        210        220        230        240
GAGTGTGATT GTTCCTCATC GCAGCATAAA CAAGTCTCTA CATCAGTGCC AGTTGCAGCC
       250        260        270        280        290        300
ACAGCCACAA CCTTTGCGTC ACAGACCTCT CAGACGCCTC GGAAATAAAC ATCGGATCAA
       310        320        330        340        350        360
TCCTTGCTGG GCTCACTTTA TCCCCGACGC AGCAGCGCGT AGCCGAGTTA CTGCGCAGGC
       370        380        390        400        410        420
ATCCGACAGA AAGCTAACAT CGATGTCAGG TCCAAGCAGA AGTGGAATTA AGGCTCAGCT
       430        440        450        460        470        480
GACAGCGGGA GAGCTTAAAT TTGCTATCTA GTGTGTGGTC CGCCGTGATC GTATAGGGGT
       490        500        510        520        530        540
TAGTACTCTG CGTTGTGGCC GCAGCAACCT CGGTTCGAAT CCGAGTCACG GCATTATCCT
       550        560        570        580        590        600
CTGGTCACTT TTTTGCTCCA CTCTCTCTCT GATGAACTCT TCCCTACAGA TCTACCCGCT
       610        620        630        640        650        660
TCCGCTCATC GTCACCCAGG AAGCGTGGGA AGTGCTTGCT CTCTCCCAAG CTGTTTTGCA
       670        680        690        700        710        720
GGAAATGGGA ATGAACCTTT AATGTCTTTG AATCCATCCT GCTGCAGCGG GCTGTCAGCA
       730        740        750        760        770        780
GTCAGCATTC ACCTTCTTCA TATTTGTATG CATATTGTAA TAAAACTACC GAAGCATCTT
       790        800        810        820        830        840
ACAATAAAAT GGTTTTGAAA AGTCAACACC TGGACCAGGT TACTGTGAAA TTTCCTCATC
       850        860        870        880        890        900
CGTCTGTGAG GGGAGGGGTG GAGAGGGAGG AGGGACAGGG AACCATGGTG CATCCTAGAA
       910        920        930        940        950        960
GGTCAGAGGA AAATTTTCAG GAATTGAAGG AACTTGGCTT TACCAGGCTT TCAGGGCAAG
       970        980        990       1000       1010       1020
TCCTCCCACC CACCTTAAAA AGGTGCGGCC CACTGGCTTT TTCCTGTTTG CTTGTTTTTG
      1030       1040       1050       1060       1070       1080
ACAGGGTCTC ACGCTGTCCT GTAACTTGAA CTCCTGATCC TCCTGCCTCA GACCCTGAGT
      1090       1100       1110       1120       1130       1140
ACTAGGATTA CAGACATTAG CTAACAAGCC AGGCTGAGTT TTACTCCTTG ATTCTCATGA
      1150       1160       1170       1180       1190       1200
GTGAATGTGC CTCCGCTGGC CCACCAGGTT TTTCCATAGC TTTCGTGCCT TGGGAATCTG
      1210       1220       1230       1240       1250       1260
TCCACTTTGG TGTGGTTATC ACTCCTTCGT TTGATTTCCT TTTGGTTTGA GTGGGAGGGC
      1270       1280       1290       1300          3         13
GATATGTATG TGTTCTATCA CACTGGATTC CATCTTGGGG GTCTAGA
```

FIGURE 24 DNA Subsequence. Part of the DNA sequence of a mouse. The sequence shown includes a gene that codes for a form of RNA.

lems, as they allow for the accurate recording and storage of the data.

A more fundamental use of computers is in the data processing itself. This can be illustrated by one of the methods devised by the British biologist F. Sanger, known as "shotgun sequencing." This process involves what is essentially a problem in cryptography. Scientists are trying to reconstruct a long message, the DNA strand, from a large number of short overlapping segments, in no definite order, which they have obtained by actually cutting the strand at selected places [Figure 25]. The individual segments cannot be more than a few hundred bases long. If the original message is much longer than this, there will be a large number of segments that need to be fitted into the proper order. Human

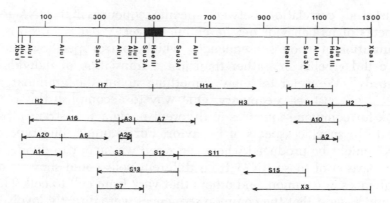

FIGURE 25   A Method for DNA Sequencing. The long bar at the top represents the entire sequence given in Figure 24. The small bars below indicate subsequences that were fitted together by examining their regions of overlap.

beings could try to do this unaided, but finding the right sequence could involve making tens of thousands of comparisons with no errors. Simple computer programs have been written to obtain sequences containing up to tens of thousands of bases, and these techniques can probably be extended to the point where the total DNA sequence of a bacterium, or of a mammalian chromosome, can be obtained. We would then know the complete genetic information of a living organism.

Such information would by no means be the end of the story, especially for higher organisms. We would know the organism's genetic code, and therefore all of the proteins that this organism can produce. But this information would not be enough to understand all of the behavior of the organism. It would correspond to having the text of a book, but without much understanding of the language in which the text is written.

There are several reasons for this. We have seen that much of the DNA sequence does not code for proteins. Some of it is used to produce other important biochemicals, such as RNA. Other parts control the expression of the coding portions, and cannot be understood just by knowing the code for proteins. Furthermore, in an actual organism, the proteins are produced in specific combinations and under specific conditions, all interlinked in a complicated way.

In order to understand how an organism functions, biologists must study correlations between specific sequences in its DNA and aspects of its behavior on the cellular level. What we need is a translation of whole subsequences, which control specific aspects of cellular behavior, rather than just a translation of individual "words." We need to know something about the grammar of DNA, not just its vocabulary. One way to accomplish this is to look for common sequences in different nucleic acid strands that relate to specific aspects of behavior. For example, one type of RNA might be produced whenever a cell was about to divide. A comparison of these RNAs from different cells could show some sequences in common, and others that vary from cell to cell. This would suggest that the common sequences were directly involved in cell division. Studies of this nature have begun, and will eventually provide us with an understanding of the sentences and paragraphs of the genetic code, instead of just the letter by letter understanding that we have now.

Another approach involves the study of the connection between the linear sequence of bases and the ways in which nucleic acid chains fold themselves in two and three dimensions. As with proteins, it is likely that the way in which nucleic acids function biologically is closely connected with these multidimensional aspects of their structure. We do not yet know how, or even whether, the sequence of bases determines these higher structures of nucleic acids, but studies are underway. Other questions, such as what information in DNA controls the transcription of DNA into RNA also can be studied by analyzing sequences of DNA and comparing their form in different cells and different species.

Computers will play important roles in such studies. For example, if several nucleic acid sequences, each containing hundreds or thousands of bases, are to be compared for common segments, the project could involve millions of individual comparisons. Computers are clearly superior to human beings in such tasks, and using them for this purpose will allow us to discover correlations in the nucleic acid sequences that might otherwise remain hidden.

Data banks are being compiled from the known sequences of nucleic acids. It should be possible to automatically compare new sequences with those already in the data banks, to see what kinds of similarities exist between them. Ultimately, such studies with computers will help us understand how living things translate the genetic information in their cells into the specific activities that we call life.

How Computers Affect the Way Science Is Done

The role that computers now play in theoretical and experimental science has changed the talents needed for effective research. It has also brought the two realms closer together. Experimental scientists need to become more expert in the manipulation of symbols and in the logical analysis that is required to write effective computer programs. Theorists—whose appetite for computations is often limited both by the funds needed to buy computer time and by the capability of existing computers to perform calculations—have taken to designing and in some cases actually building computer systems tailored to their specific needs. These trends will lead to a blurring of the sharp division between theoretical and experimental scientists that has marked twentieth-century physics.

Today, graduate students both in experimental and theoretical physics spend most of their time writing and debugging computer programs rather than devising new instruments for making observations or finding new ways of solving equations analytically. When these students become leaders of research groups, they will naturally think of using computers in most research problems.

Computers will become the norm in all areas of science in which large amounts of data can be obtained. Even fields such as evolutionary biology will benefit from the availability of such data. Researchers in molecular evolution, who study the changes in the nucleic acid sequence of different organisms over time, have already begun to make good use of data processing.

This change in the way that data is handled could have interesting consequences. In the sixteenth century, Francis Bacon described a model for scientific research that consisted of gathering data and then searching for patterns within it that could be regarded as general laws. This Baconian model has not really been followed by scientists because there was no straightforward method of finding the patterns in the data. Computers, which can analyze vast amounts of data and search for patterns, may direct science into a more Baconian mold than would have once seemed plausible. Whether this approach to scientific research will work depends on the interrelations between the ideas of science and the structure of the human mind, a question that has never truly been answered. We do not know to what extent the ideas that we use to describe the world are the result of human idiosyncrasies, as op-posed to being in the world itself. It would be fitting if the answer comes not from philosophical speculation but from the empirical methods of science itself.

Future Computers

No discussion of the role of computers in future science would be complete without some consideration of the further evolution of computers themselves. Computer capabilities are evolving very rapidly. One result of this evolution will be that computers will not only become better tools to aid human thought but also partners in human thought.

One approach to making computers into more effective tools is through the development of better hardware to perform the operations. Such hardware will contain more individual elements integrated with one another, which would mean that more elements could rapidly influence one another. Such a computer could better perform such operations as the manipulation of mathematical symbols. At present these operations need to be programmed afterward, and so are done less effectively.

The size of individual computer elements will continue to get smaller, and this will allow for even greater capacities for speed and integration—though ultimately there are limitations. We know of no way by which the individual elements can be closer together than atoms in a solid body, and no way in which information can be transmitted between these elements faster than the speed of light. However, these limits would still allow for the construction of computers that are millions of times more effective than any attained thus far. Developing computers that use individual atoms as elements will itself require a novel technology.

Improvements in computer speed and integration, and concomitant decreases in computer size, will make future computers considerably more useful in their scientific applications. Current computer capabilities become rapidly saturated by even modest calculational efforts, especially when numerical calculations need to be done in three or more dimensions. Increased speed, and the decreased cost that should result from decreased size, will make possible calculations and data analysis that are substantially more sophisticated than those done at present.

Faster and cheaper computers will also accelerate the trend toward decentralization in the use of computers. In many scientific applications centralized mainframe computer installations have been replaced by individualized microcomputers. As computers become cheaper, more and more scientists will rely on their own dedicated computers, rather than on a small share of a central computer. This will allow for a greater customizing of computers for the purposes of individual scientists, a development that will lead to even more rapid evolution of computer capabilities. This trend toward decentralization will also help eliminate one of the main problems in the use of computers, the inaccessibility of the central computer, either because too many people want to use it at the same time or because it has temporarily broken down.

Another form of evolution of computers involves developing novel methods for the logic by which computers solve problems. One method that is increasingly being applied is known as parallel processing. It involves dividing up a problem into many identical parts, each of which is done at the same time (in parallel) on many small computers, all of which are controlled by a central unit. Arrays of parallel processors have already been built for some applications in particle physics. They can do specific calculations much faster than the best existing general-purpose "supercomputers," and at much lower cost. Several of my colleagues in the physics department at Columbia University are using this method to study the interactions of quarks and gluons in discrete space-time [Figure 26].

An advance in the logic of computer applications has one important advantage over new computer hardware: It is more directly under the control of the scientists themselves, and therefore it appears to be less of an intrusion into their research. Also, a scientist who develops a new and inexpensive way of using existing computers is less likely to infringe on the resources of his colleagues than one who insists on the purchase of new and expensive hardware. Both of these factors will help to smooth the path toward the general acceptance of computers in science, even in places where they are not now used.

One of the most intriguing possibilities is that computers will someday be developed that equal or even surpass human capabilities for creating new scientific ideas. This possibility is an outgrowth of the field of research known as artificial intelligence. Workers in this field attempt to construct and program computers to duplicate specific human intellectual abilities, such as proving mathematical theorems or diagnosing diseases.

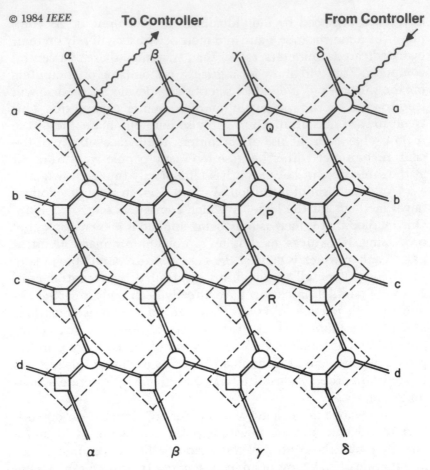

FIGURE 26   Blueprint for a Parallel Processor. Each circle and square surrounded by a dotted line represents an independently functioning microprocessor and memory unit. The lines joining them represent the flow of information between these units.

There have been two somewhat distinct aims of this research. One aim is to gain useful new perspectives on human thought by having another example of thought to study, thought with an electronic rather than a neuronic manifestation. The other aim, from which the main contribution of the field to future science is likely to come, is the development of what I call "really smart computers," whose intellectual capabilities are comparable to or greater than humans'.

The level of computer intelligence that has been achieved

thus far is greater than was expected by many critics of the field, but less than the predictions of some of those working in it. Computers have been constructed, or instructed, to carry out several activities on levels comparable to what humans can do. Some of these activities, such as playing chess, are regarded as an indication of high intelligence when humans perform them. There are other human activities, such as translating accurately from one language to another, which computers have not yet been able to do very successfully. We are better at getting computers to do those things that humans learn relatively late in life, such as playing chess, than those things we learn in early life, such as understanding speech or recognizing patterns. Perhaps the latter abilities are intrinsic to the human brain, in which case we would have to construct them directly into computer hardware rather than programming them in computer software. Alternatively, it may be that we understand less about those abilities that develop early, and so are unable to duplicate them adequately.

I think that once we understand any intellectual activity well enough to describe clearly what it accomplishes, then eventually we can teach computers to do it. Perhaps no single set of computer hardware will be able to perform every "intellectual" function—but then, no single human can perform at a high level all of the intellectual activities that are within the capability of at least some members of our species. Furthermore, for activities where there are different levels of excellence, such as playing chess, it might be possible for computers to eventually exceed the best human achievements. Why? For one thing, computer performance can be made to build more directly on past achievement than does human performance. While it is true that the best chess players use the games of their predecessors to help them, computers could do the same thing with more reliability and faster access. And while there may be no way of duplicating the quirks of brain function that help produce human chess champions, it is simple to incorporate into future computers any improvement in computer design that allows for better chess play.

Such analysis has convinced many scientists that most of the specific intellectual functions that human scientists perform could eventually be performed more effectively by computers. If this prediction turns out to be correct, the future content of science and the lives of future scientists will change drastically. Science is done by scientists, and supported by society, for two purposes: to understand the universe and to ameliorate the human condition through technology. Really smart computers would have different effects on these two aims.

Scientists are interested in understanding the universe in terms that make sense to human beings. For this purpose, the existence of computers that can understand things in their own ways could be irrelevant. Actually, few of the leading research scientists that I know avidly follow even the work of other scientists. They may try to use the general ideas of others but prefer to work out the details themselves. For many active research scientists, the quest for understanding is personal, and not just a part of some collective process. It is a paradox of science that while such understanding requires general acceptance by a community working on the same problems, the understanding resides ultimately in the minds of individual scientists, where it only comes to exist after it has been expressed in terms that are personal.

Consequently, really smart computers used in scientific research will not become "oracular," able to ponder the universe on their own and then to answer any questions that scientists ask them. If such machines were developed, they would be irrelevant to the type of personal scientific understanding that scientists seek. It is possible that we might want to develop such computers anyway, in connection with the technological applications of science. There, understanding is subordinate to the aim of modifying the universe. A computer that could tell us how to build a better spaceship would be very useful, even if it couldn't explain to us the view of the universe that led it to its conclusions.

A more likely application for really smart computers in science will be as partners to individual scientists in their research. In order for this development to take place, computers will have to become much more capable of receiving and processing data directly from humans. This could be done either by teaching computers to understand human speech, or perhaps by giving computers direct access to the human brain through some kind of electronic hookup. It does not seem too difficult to accomplish such communication, especially if the computer need only interact with one person.

I imagine a future in which early in the training of a scientist, the computer will act as a teacher, and later as a research collaborator. This computer would have rapid access to central computers with immense information and data-processing capabilities, which it would make available to the human. Communication between computer and human would be as rapid as between two people, possibly much more rapid, if the computer has direct access to the brain. Such close communication and recognition by the computer of the modes of thought of an individual is currently beyond the ability of computers, but probably not for very much longer.

Most scientists would probably enjoy having such a computer partner, whose thought processes were, from long common experience, very similar to their own. It would be similar to having a long-time human collaborator, but without the personal problems that often arise in such cases. The results of such collaborations between humans and machines could transcend the results of independent efforts of the two partners.

If intelligent computers can serve scientists as reliable partners in thought, this would be a greater return on our intellectual investment in artificial intelligence research than almost any specific scientific discovery that computers would make on their own. Such a partnership is perhaps the greatest promise that artificial intelligence research holds for scientists—and for the future conduct of science.

Part 4
Science in the Twenty-first Century

How Science Will Shape the Future: The Technological Adventure

8

There are at least three types of future technology. One involves the application of presently known science, and only requires engineering development or time and money to implement it. A recent example was the building of the first nuclear reactor, once the details of nuclear fission had been understood. A present example is the building of nuclear fusion reactors.

A second type involves setting our sights on technologies that can only be accomplished through new scientific discoveries. Finding a way to stop and reverse the process of human aging would be an example. A past example was the desire to cure infectious diseases such as tuberculosis, which led to the development of antibiotic therapy. Charles Babbage's proposal in the nineteenth century to build a general-purpose computer was another example. Although Babbage had the right idea about how such a computer could be designed, implementing his dream required the development of electronics. Anticipating this type of technology is difficult, and requires some imagination.

Finally, there is technology that is so revolutionary it cannot be imagined at all. There was a time, for example, when travel to other planets was literally inconceivable. Not until the time of Galileo and Kepler was it realized that some of those twinkling lights in the night sky were actually worlds like our own.

In my discussion here, I will omit the first type of technology; it has been addressed by many authors and also overlaps little with other sections of this book. Most of what I have to say will be about the second type of future technology, guided by my own educated guesses about future science. Little will be said about the third type of technology, for obvious reasons.

What is the relationship between science and technology? As we have seen, it is not uncommon for a specific scientific investigation to be motivated by the wish to develop a particular technology. For example, much current research in cellular biology is an outgrowth of the wish to find a cure for cancer and of the decision by various funding agencies to support research that might lead to this goal.

However, technology can also progress independently of science. That is, useful discoveries are made even when there is little or no understanding of how these discoveries work. This has been the case for many discoveries in medical technology, such as analgesics. It is still not well understood how a substance like aspirin acts to relieve pain—though such gaps in understanding often spur scientists to new areas of research. This pattern, in which technological developments occur independent of scientific understanding, is becoming less common as science comes to encompass an increasing portion of the phenomena directly concerned with human life. However, "autonomous" technological discovery will continue to be responsible for some of the most interesting future technologies.

In describing possible future technologies, I am not necessarily implying that I "approve" of them. Some could lead to sweeping changes in human life—and even in what human beings fundamentally are. Yet it is exactly because these technologies may change our lives so dramatically that it is imperative to imagine their effects now.

Physical Technologies

Most technology in the modern world has involved the application of the physical rather than the biological sciences, and so it is the nonliving world—the world of objects and "hardware"— that has been most transformed. Examples of this kind of technology are computers and jet planes. Though it seems likely that biological or psychological technologies will play the greater roles in the future, there are still many interesting possibilities for physical technology as the physical sciences continue to make new discoveries. Two of these possibilities are the development of means for exploring presently inaccessible environments, and the building of structures in which individual molecules are arranged according to an overall plan.

We tend to think of earth's surface as containing a wide variety of environments, but actually the range of temperture, pressure, radiation, or most other physical parameters is not very large. This is the narrow environmental range in which unprotected humans can live and function. But most environments in the universe, and some in or near the earth itself, are very different. In order to explore these hostile environments in person, we need substantial protection; we must take our familiar environment along with us, in the form of a spaceship or a bathysphere. Alternatively, we can send our surrogates—automated equipment—to do the exploring for us.

This "surrogate" approach has already extended the range of environments that we can explore to include the bottom of our oceans, the surface of Mars, and interplanetary space. But there remain interesting nearby environments, such as the deep interior of the earth, that even our instrument proxies cannot yet explore. An important challenge to future technology is to develop methods for such exploration.

The type of environments that we presently have the most trouble exploring are those in which both the temperature and pressure are high. We can to some extent explore high-pressure, low-temperature environments, such as the bottom of our oceans, both with automated probes and by humans in protected environments. We can also explore parts of interplanetary space, where temperatures are high but pressures low. But we have not yet been able to study for more than a few minutes an environment such as the surface of Venus, where both high temperatures and high pressures prevail. In such environments, there are constant collisions between the atoms of the environment and those of any probe that we introduce to study it. These collisions act to raise the temperature of the probe so that it is equal to that of the environment. Also, high temperatures and pressures tend to speed up the rates of chemical reactions between substances in the environment and the materials of which the probe is made. As a result, the probe soon changes its physical and chemical form and can no longer function.

Scientists have tried to find ways to protect the working part of the probe from this type of environment, but with mixed success. Often there are many different forces acting to destroy the probe, and it is difficult to anticipate all of them, let alone protect against the ones we know. This was the problem with some of the Soviet probes sent to Venus over the last decade. Each successive probe was engineered to withstand the factor that had destroyed the previous one, but each new model quickly succumbed to some other factor that hadn't had time to act before.

Though it is likely that we will soon construct a sturdier probe to study Venus and other similar environments, there are other environments, such as the interior of the sun, that we cannot hope to explore with any device made of ordinary matter. However, the universe contains types of matter other than those found on earth, and it is possible that we will find a way to use them to protect our instruments—and maybe even ourselves—from destruction in extreme environments.

For example, inside of a type of aged star known as a white dwarf, a form of matter is found whose density is as much as a million times greater than that of ordinary matter. Matter of still higher density is found on and inside of neutron stars. These and other known types of dense matter would be invulnerable even in some of the most unpleasant environments to be found in our solar system.

But there is a catch to all this. As far as we know, this form of matter can only be made, and is only stable, under the conditions of immense gravity that occur in collapsed stars, such as white dwarfs. Even if we could somehow transport some of this matter to earth, it is likely to become unstable, and change into some form of ordinary matter even before it gets here. Yet the prospect of someday using collapsed matter may not be entirely bleak. Collapsed matter either contains no electrons or contains electrons that are much more tightly bound than those in ordinary matter. It is this tight binding that gives the matter its high density and resistance to destruction. It may be possible to induce this strong binding and to stabilize this form of matter through environmental factors other than gravity, such as with intense magnetic forces.

The known types of very dense matter are made of ordinary subatomic particles—neutrons, protons and electrons. Other very massive, stable subatomic particles may also exist. If so, it is possible that such particles could be used to produce very dense forms of matter that would be stable on earth. Any form of matter not containing electrons would be extremely dense, and might serve this purpose.

Absolute stability is not necessary in order for a form of matter to be usable. Carbon in its form of diamond is unstable, and eventually reverts to graphite. But this takes a very long time, and during this time, the diamond can be used. Perhaps the same will be true for some type of collapsed matter.

If collapsed matter can be produced and preserved at least temporarily, it would have many applications in addition to its use as a form of protection for exploring hostile environments. Possibly such matter could be used as shielding against the effects of nuclear explosions; it therefore seems clear that studies of how to achieve the use of collapsed matter need to be pursued vigorously.

Another approach to the problem of exploring hostile environments involves designing the instruments so that they are adapted to the alien environment rather than to the earth's environment. Every environment, no matter how different from our own, contains forms of matter that are relatively stable. For example, deep inside the sun the stable form of matter is plasma, in which the nuclei and electrons have become separated. It seems plausible that for each environment, we can design instruments that utilize forms of matter stable within it. The problem of how to observe the things that interest us with the use of such locally adapted instruments will require study, but I do not think it insoluble.

Molecular Technology

We usually think of physical technology as performed on a grand scale, involving the construction of bridges, spaceships, pyramids, and particle accelerators. But we are beginning to explore another frontier for physical technology that may be of equal importance in the future. In a speech given in 1959, entitled "There's Plenty of Room at the Bottom," the physicist Richard Feynman suggested a novel technology, in which we will manipulate much smaller units than we have been able to thus far. In the years since his speech some progress has been made in this direction, but there is still a long way to go before we reach "the bottom."

Until now human technology has involved the arrangements

of objects quite large in comparison to the individual atoms that compose matter. Although chemical technologists have been successful in producing new types of molecules, they have not been able to arrange molecules into structures so that the position of each molecule is determined by an overall plan. The small size of individual atoms, and the large number of them in an object of ordinary size, have made engineering on the molecular level seem virtually impossible. Yet such molecular engineering is routinely carried on by even the lowliest of living things during the process of development. We do not yet have the proper techniques to accomplish it. But this situation is changing, and, especially through advances in our capability of using coherent, short-wavelength radiation, we too will be able to match the accomplishments of the cells that compose us, and do molecular engineering.

These molecular structures will be much more complex than anything that human technology has thus far achieved. The information contained in the structures will rival and eventually surpass that in the genetic material of even the most complex living organism. Furthermore, the number of structural elements that have individual functions will surpass even that in biological systems. Through this "nanoengineering" (a term I use to distinguish this technology from current microengineering), which works with combinations of billions of molecules, we should be able to produce systems with trillions of structural elements (whereas microengineering involves merely thousands or millions of structural elements). As a general rule, the more working elements in a system, the more complex the tasks it can perform.

Nanoengineering will involve the fabrication of minute structures of astounding complexity. Suppose that the structure was a lattice, in which each lattice site could be occupied by one of ten different atoms. The description of the lattice would list the sites and their associated atoms [Figure 27]. If the lattice were a cube with an edge five micrometers long—about the size of an average bacterium—and the space between atoms were about one-half nanometer, then the lattice would contain a trillion sites. The description would require several trillion "words!" This is approximately as many words as are in all the books ever written, and the description would contain about as much information as does the genetic material of twenty thousand people. The construction of the description alone would be a formidable job. Yet we are already beginning to manipulate and store such quantities of in-

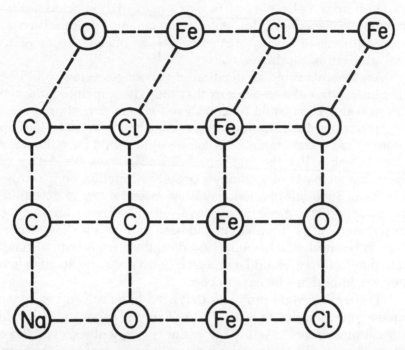

FIGURE 27 A partial blueprint for a Hypothetical Nanoengineered Structure. The blueprint indicates which of several atoms is to be placed in various locations of a cubical lattice.

formation in our computer technology, so this will not be a major hurdle to overcome in the development of molecular structures.

There are several technological needs that could be satisfied through nanoengineering. One is the development of new materials, with such desirable properties as extreme strength. Existing materials break under relatively low strains because of small regions inside the material where the component atoms are improperly aligned. If we can control the arrangement of the individual atoms, we could eliminate such "dislocations," and increase the strength of common materials by a factor of many thousands. The availability of such superstrong materials would have major applications in structural engineering. If the cables supporting suspension bridges were made of such materials, they would need to be only millimeters thick instead of meters thick as they are now. Cables could be thousands of miles long without breaking under their own weight, and could possibly be used as a means of transportation from a planetary surface to a space station or even from the earth to the moon. Also, superstrong common materials could be substituted for critical materials such as manganese or platinum, alleviating shortages.

Another plausible application of nanoengineering would be the construction of microsensors that could be implanted into the human body. They could remain there for long periods, monitoring various physiological functions from subcellular molecules up through tissues and organs. This information would have immense scientific value. For the first time, we would know the details of how different parts of a complex organism function on a continuous basis. Such information may be an essential key in determining some of the mechanisms involved in growth and aging. Furthermore, after information had been gathered for some time, it could be used as a baseline, and deviations from normal functions due to illness could be detected at a very early stage, when they would perhaps be easy to correct.

There are several problems that need to be solved before we can use microsensors. Few people are likely to submit to multiple surgical procedures to satisfy the curiosity of scientists, even if the information to be obtained promises great benefits. However, techniques may be found through which sensors could be targeted on specific tissues; after being ingested, inhaled, or injected, they would be transported by ordinary body mechanisms to the desired location, where they would implant themselves automatically. There are the connected problems of possible damage to the body by the presence of the sensor, and action by the body's immune system to expel the sensor—but these, too, will probably not be insurmountable.

Another serious problem may be getting the information obtained by the sensors to the outside environment, where it can be evaluated. It may be possible to devise sensors that broadcast their information as impulses of microwave radiation, to which living material is relatively transparent. This radiation could then be detected outside the body, and decoded there.

Nanoengineering is a task for the future, but we have already made progress in bridging the engineering gap between ourselves and the molecules. Microengineering is already here in the design of microcomputers. In such designs, we regularly produce structures in which individual elements whose size is a micrometer or less are placed into a desired arrangement [Figure 28]. This is

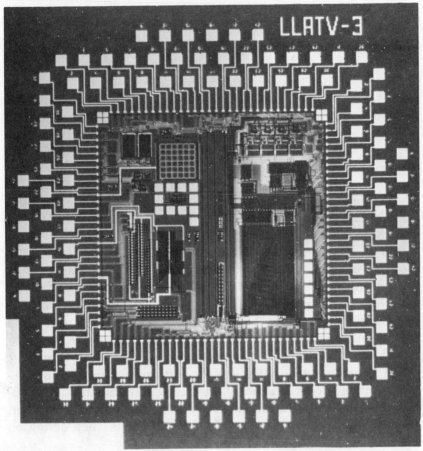

FIGURE 28 Microengineering Structure. A computer chip containing logic switches and other circuits. The whole chip is 4 millimeters across, and some of the channels through which electrons flow are a micrometer long.

done by a number of techniques. In one type, known as lithography, a pattern is impressed onto a surface by selectively shielding parts of the surface with an opaque mask. Radiation is then beamed through gaps in the mask to expose a sensitive material on the surface, and then the exposed material is removed by chemical or other means.

The accuracy with which this can be done depends on a number of factors, including the wavelength of the radiation used. With visible or ultraviolet light, patterns can be made to an accuracy of about a micrometer. To make more accurate patterns, as well as to make the mask itself, X-rays and electron beams can be used. Extending the techniques of lithography by use of intense beams of short-wavelength radiation promises to increase our accuracy to the point where patterns that are smaller than viruses, just a few nanometers long, can be produced.

There are several distinct approaches to molecular engineering. One would involve the extension of methods already used in microengineering; the nanolithography discussed above is but one technique. The "extension" approach could involve the use of external probes, such as electron beams or X-ray beams, to shape the unformed material into the desired molecular arrangements. (Visible light beams cannot be used for this purpose because the wavelength is too long. It would be like trying to build a watch by using a bulldozer.)

It may be possible to extend some of the methods using the short-wavelength coherent radiation described in Chapter 5 to "nanofabrication" as well as to seeing what we are doing in the nanoworld, since microholography can be used to demagnify patterns as well as magnify them. Such a demagnified pattern could be imprinted into a prearranged layer of material by the techniques described above. It should even be possible to transport atoms selectively to particular sites on a surface by accelerating them with beams of coherent short-wavelength radiation. This approach will require the building up of one thin layer after another, because such external probes are most effective when acting on the surfaces of objects.

In his speech on nanoengineering, Feynman suggested yet another approach: In a series of steps, we could construct small tools that could be used to manipulate still smaller objects, out of which still smaller tools could be made, until we reach the molecular level. This particular approach may become more feasible if some of the microtools have computers built into them to control the manipulations of molecules. Human reaction times may be too slow to control molecular processes directly.

A third approach would be based on the procedures through which bacteria and other living things produce the protein molecules that they use in their life processes. They produce the protein as a long chain of amino acids which then folds itself into the required three-dimensional arrangement. This folding takes place not through any action by the bacteria, but as a consequence of the operations of the laws of physics on the protein chain in the cellular environment in which it was produced. Proteins are not the only class of substances which, once they are produced with a suitable molecular composition, "automatically" configure themselves in the correct way to perform some function. Some components of nonliving things, such as the atoms in a metallic alloy, do this as well.

We do not yet know enough about physical processes on the level of large molecules to be able to predict precisely the physical arrangements they will assume, but this problem is not impossible to solve. Once we understand how and why biomolecules configure themselves in three dimensions, we should be able to extend this understanding to molecules of our own creation. Consequently, we may be able to avoid some of the problems involved in specifying where every one of trillions of atoms needs to be placed.

It is even possible that variant forms of biomolecules themselves could be useful to us. Polymers can exist in an immense number of types, only a few of which play biological roles in any life form. Other polymers, with highly specific chains of monomers, might be used for nonbiological purposes. For example, it has been suggested that some organic polymers might exhibit very high electrical conductivity, and so be replacements for metallic conductors. If organic polymers do turn out to have such valuable uses, we could probably "train" living things, such as bacteria, to produce the novel organic polymers for us, without our having to do molecular engineering.

Nanoengineering holds two promises for future technology. One is the construction of objects of a previously unimagined complexity, with an extended range of capabilities arising from that complexity. The other is the redesign of materials through control of individual atomic placements. These promises are likely to be realized in diverse ways, and shape our physical technology into forms that cannot now be foreseen.

Biotechnology

The most important type of future technology will involve the manipulation of living matter, especially human cells, rather than of inorganic matter, as is done by physical technology. It is appropriate to refer to this new technology as biotechnology, rather than to name it after one of its specific manifestations, such as genetic engineering or recombinant DNA.

Because the science of biology does not yet have as deep an

understanding of its subject matter as do physics and chemistry, biotechnology is at a relatively primitive stage. Yet its future will include the capability of modifying the most fundamental aspects of human life, such as aging, sexuality, inborn inequalities, and sociopathic behavior. Among the future developments in physical technology, probably only those in artificial intelligence will rival biotechnology in the scope of its possible influence on human life.

One important type of biotechnology is genetic engineering, the alteration of genetic information in order to change some cellular or bodily function. For example, some diseases appear to involve mutations in a single gene. There is every reason to believe that, as techniques advance, we will be able to return the gene to its normal form and eliminate the malfunction involved in the disease.

Within the framework of genetic engineering, there is a further distinction: between changes in the germ cells, which would affect only future generations, and changes in somatic cells, which would affect the living organism. The second type of change is likely to be more difficult, as it would involve changing a large number of cells while keeping the organism viable.

Inducing change via the germ cell is not new; selective breeding of plants and animals has been done for millennia. This form of technology has enabled us to explore the wide range of natural variation in the gene pool of any species, and produce living things with those characteristics that we prize. Some of the variations that can be expressed through selective breeding are quite remarkable. For example, dogs range in weight from 1 kilogram to 100 kilograms and are still members of one species.

However, there are limits to what can be accomplished by selective breeding, which merely enhances characteristics already present in a species. If a characteristic requires proteins that are coded for by nucleic acid sequences not found in the genes of any members of a species, then this characteristic usually cannot be introduced through selective breeding. For example, human beings cannot synthesize eight of the amino acids that occur in the proteins in our cells. In order to live, we must consume proteins that contain these amino acids. Suppose we decided that it was desirable to produce human beings that could synthesize these amino acids in their own bodies. Such people could live on a much simpler diet than we can. In order to do this, it would be necessary to use methods other than selective breeding.

Techniques already exist for altering specific parts of the DNA sequence, at least in laboratory conditions. Substances known as restriction enzymes can chemically alter the nucleic acid so that the sequence is cut at a point along it, and other enzymes can introduce a specific new subsequence into the nucleic acid chain where the cut was made. The new sequence would then result in an altered function of the cell containing it. It would be easiest to do this in a germ cell or a fertilized egg, which could then be implanted and develop normally.

It would be more difficult to accomplish this in the many cells of some organ of a living multicellular creature—in effect changing its ongoing cellular functioning—but we should be able to do this in the future. Prospective micromachines and nanomachines, implanted into the organism, would be the proper size to manipulate individual cells and parts of cells, and so could be used to carry out subcellular reconstructions without substantially disturbing the whole organism. Alternatively, we may be able to get viruses to selectively transport new genetic material into the proper cells in living organisms. Viruses already do this for their own reproductive purposes, and we should be able to induce them to use their technique for our purposes.

However, even when such techniques exist, there will remain a substantial barrier to many types of effective genetic biotechnology. This barrier arises because of our limited understanding of the complex relation in higher organisms between the message that is encoded in the DNA sequence and the way in which this message is represented in the function of the cell and in the behavior of the organism. We do not even understand this on the level of individual cells. It has been discovered recently that the DNA in some human cancer cells differs from that in a corresponding normal cell by only a single base at one point in the DNA strand. This is about the simplest possible difference that could be imagined. However, no one has yet understood how this small variation manifests itself in the many differences between normal cells and cancer cells, such as the limited division potential of the former compared to the unlimited division potential of the latter.

There are many possible forms of biotechnology; their implications are so far reaching that I cannot hope to cover them in depth here. In the very near future, the forms of human biotechnology that are most likely to be developed involve the cure or elimination of genetic diseases, and the control of such reproductive matters as the sex and number of offspring. Other forms of human biotechnology are probably more distant, but I want to briefly discuss two of them—growing new organs and control over aging.

A method for growing organs, genetically identical or very similar to a person's own, could ensure successful transplants when the original was destroyed by disease or injury. These transplants would not be subject to the immune reaction that interferes with present transplants of foreign organs. The development of such a technique would also eliminate the problem of scarcity of organs for transplants.

Once biologists understand the process of development as it takes place normally, it should be possible to induce it artificially, using cells taken from an individual's body, which contain all the genetic information originally used to produce the organ. It might be necessary to cause some cells to revert to the undifferentiated state that they had in early embryogenesis; or we may find that cells capable of redevelopment occur normally in adults. Perhaps a few cells of different types will routinely be taken from everyone during infancy or gestation and preserved at low temperature until they are needed to grow new organs. This type of preservation can already be done with small numbers of cells.

The process of growing new organs might take place within the person's body or in some external artificial medium. Within the body, there are natural methods for nourishing and integrating a growing organ. It might, however, be difficult to avoid interference with the functions of the other organs while a new heart or liver was growing within the body. Using an external medium would circumvent that problem, although it might make it more difficult to ensure that the organ developed normally. Yet even now, organs that have been removed from a body can be nourished and maintained for short times with specialized devices, and these could probably be adapted to allow growing organs to develop. It would, of course, be necessary to surgically implant the mature organ, but this should pose no great problem.

It might even be possible to improve upon the original organ and still avoid rejection. For example, if the original organ had developed a tumor because of some genetic or other identifiable defect, the defect could be eliminated in the new organ before it was implanted. This type of replacement therapy could ameliorate the negative consequences of the aging process. A fifty-year-old man might be as healthy as he was at twenty if his worn organs were replaced by new improved copies.

The possibility of organ replacement does not extend to the central nervous system. Here the most likely technique would be to induce regeneration of damaged parts of the brain and spinal cord while preserving both the memories stored in the undamaged parts of the brain and the exquisitely complex connections that characterize the mature nervous system. We may not need to understand all aspects of development to achieve this technology of regeneration. Peripheral nerve regeneration already occurs spontaneously in humans, and limb regeneration takes place in amphibians. Probably what is needed to make central nervous system and limb regeneration possible in humans is a better understanding of growth mechanisms in adult organisms.

A technology for growing new organs and regenerating old ones would help alleviate many of the health problems that currently cannot be treated, such as degenerative diseases of the central nervous system. This technology would prove equally important in diseases that we can already treat, such as kidney failure, where the expense of treatment severely strains our medical services system. However, even this technology would leave untouched the problem of the loss of biological function that occurs because of the aging process.

The human life span could probably be extended somewhat through a series of organ transplants. But if our life span is to be substantially increased, we will have to find some way to slow and reverse the aging process itself. An antiaging therapy might develop out of an understanding of the causes and mechanisms of aging, or it might be developed through chance discoveries. However we come to it, some aspects of antiaging therapy can be foreseen from what we already know about aging.

As Figure 29 shows, the deleterious effects of aging begin at age forty or even earlier, so an antiaging therapy would have to begin early in life. What is taking place is an overall process, at least on the organismic level, so a true antiaging therapy would have to treat the overall cause of aging, rather than its effects on individual organs. How easy it will be to do this will depend on the actual causes and mechanisms of aging.

One consequence of the aging process is a diminished ability to resist various diseases. As a result, the death rate increases rapidly with age, and the maximum human lifetime is about 120 years. Nothing we have been able to do until now has led to any change in this maximum lifetime. By preventing or curing a number of individual diseases, we have been able to increase the average human lifetime (or life expectancy) to its present value of about seventy-five years in many countries. However, it is unlikely that this approach will extend the average lifetime much further. Because the rate of death due to many different diseases increases so rapidly with age, even if one or several of the most important present causes of death among the aged were eliminated, we would still die of other diseases that had not been eliminated.

The death rate at any age is the product of two factors. One is a vulnerability factor, which measures the number of different causes of death and which is relatively constant after age thirty. The other factor measures the result of aging. This factor alone leads to a doubling of the death rate for every seven additional years of age. From the second factor, we can infer a rate of aging, as measured by the death rate. This aging rate has changed little for as long as we have kept records. In fact, over the last decade, during which life expectancy has increased significantly, the rate of aging inferred from the death rate has actually increased slightly, the increased life expectancy coming instead from a decrease in the vulnerability factor. Unfortunately, the maximum lifetime is very insensitive to changes in the vulnerability factor.

On the other hand, a treatment that slowed down the rate of aging would almost surely lead to an increase in both the average lifetime and the maximum lifetime. For example, if the rate of aging after age thirty could be uniformly decreased by 10 percent, ten years would be added to the maximum lifetime, and several to the average lifetime. Furthermore, a person would live these extra years of life in relatively good health. It is clear that control over the aging process itself is the method we will have to use if we are to increase the human lifetime significantly.

What we can do about aging depends on how and why aging occurs. If aging is due to the malfunction of a specific organ system, such as the immune system, then understanding the mechanism of this malfunction could allow us to correct it, as we have done with various other organ malfunctions. In this case, aging control would really become a part of ordinary medical practice.

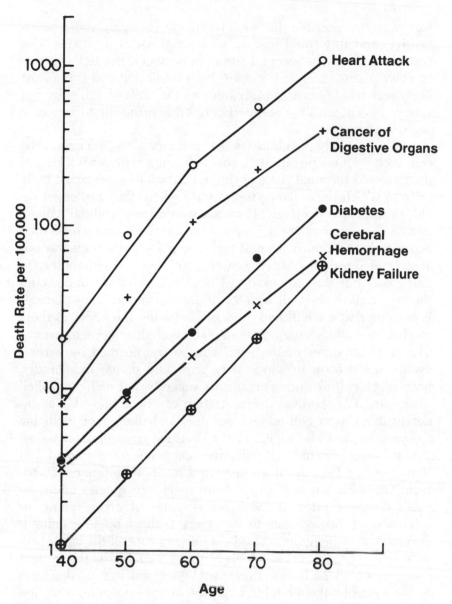

FIGURE 29 Disease Death Rate Versus Age. The rapid increase with age of the death rates for several prominent diseases is shown.

However, it is possible that such treatment for aging could be expensive, and this could lead to substantial social problems. The poor are unlikely to accept a situation in which the rich can buy off even death, while they cannot. Yet I think that this is the least likely scenario to occur, simply because the facts of aging are not very consistent with the notion that it is due primarily to a specific organ malfunction.

A more likely possibility is that primary aging occurs on the cellular level. One possibility is that cell aging represents a kind of programmed terminal stage of differentiation of some or all body cells, in which they change into a form where they no longer divide or function effectively. If so, this would make cellular aging a part of development itself. Slowing the aging process would then become part of a more general technology for influencing various aspects of development. At present, we have no capability for doing this, but the outlook is not hopeless. Once we understand the mechanisms through which cell differentiation takes place, it is possible that we will find ways of modifying the effect of these mechanisms, thus changing the rate at which development occurs. The fact that differentiated cells can be transformed by various agents into a form in which they grow and divide indefinitely, both in cell culture and in organisms, suggests that such controlled modification of terminal differentiation can be achieved. The indefatigable cancer cell may be not only a challenge for medicine to overcome but also a clue to the eventual triumph over aging.

It is also possible that cellular aging is not programmed, but that it occurs because of accumulated errors in cellular metabolism. The variation in life span from species to species might involve different rates of these errors, rates of error repair, or tolerances of the organism to the errors. If this model for aging is correct, it may be difficult to do anything to control the aging process. We may soon know which of these scenarios is the correct one, and whether a human dream that goes back in recorded form to the Epic of Gilgamesh has any prospect of becoming a reality.

Future advances in our understanding of biological processes will give us the capacity to modify our present biological functions and to introduce novel ones. Biology will no longer be destiny. We will have the power to make what we will of ourselves.

This possibility will confront us with immense ethical choices, indeed, with the choice of whether to embark at all on the reshaping of the human species. It is essential that these choices be made in the light of a full discussion of our hopes for the future of humanity and of the biosphere. We should not reject the prospect of improving our species by an appeal to catch phrases such as "No interference with nature!" or "Who will play God?"

Natural processes have no intrinsic wisdom. The nature we have inherited is the product of innumerable accidents that have taken place over four billion years. We need to understand these processes before we can sensibly alter them, but we should no more worship the existing biological character of our species than we worship the brass idols that our ancestors held holy. We should not turn our backs either on the moral questions biotechnology raises or on the exquisite opportunities it offers.

How the Future Will Shape Science: New Procedures for the Quest

Only fifty years ago the number of scientists was about a tenth as large as it is today. Currently about 250,000 physical and biological scientists are doing research in the United States, and perhaps a million throughout the world—numbers that are increasing by several percent each year. In the last few years, there has been an even larger increase in the number of scientists from less developed countries, many of whom have been trained in developed countries. Less developed countries will probably be the main source for future increases in the number of scientists.

Science has also prospered because of the immense amount of national resources devoted to scientific research. At present, expenditures on basic research amount to about $10 billion annually in the United States, and about three times as much throughout the world. Two generations ago, such expenditures were measured in thousands, rather than billions of dollars.

This large change in the scope of scientific research has resulted in qualitative changes in the working lives of scientists. These changes are most marked in physics, where expanded government expenditures on research first occurred, but they are now being felt in other fields, such as biology. The changes that have already taken place have made the day-to-day process of most scientific research a very different thing from what it was before

World War II (which marks the dividing line between the old and new eras of scientific support). The process of change is continuing and it will make the working lives of future scientists different from those of present scientists.

Group Science

One major change from the recent past in the way scientific research is done is that now a single project often involves a large number of collaborating scientists, sometimes in the hundreds. These collaborators may be based at laboratories scattered around the world and may never meet as a group. This situation is most pronounced in the experimental physics of subatomic particles. But even in some areas of theoretical science, articles written by four or more authors are not uncommon. This is in contrast to the situation not long ago, when a piece of scientific research was largely the work of an individual scientist. Many scientists rarely met with other scientists, except at conferences or during occasional visits to other laboratories.

To some extent, this multiplication of authorship is the result of real needs within science. For example, if an experiment involves the analysis of a large amount of data, it is often necessary to let several groups analyze parts of it at the same time. Also, some experiments involve so many different scientific tasks that it is reasonable to divide them up among many groups, each of which can specialize in one task. There is little doubt that many experiments in particle physics would be difficult or impossible to do without such a subdivision of tasks. However, there are other reasons for collaboration, reasons that do not always serve to advance scientific research.

In some fields of scientific experiment, such as particle physics and radioastronomy, the possibility of doing experiments is limited by access to complex instruments, which only a small number of researchers can use at any one time. As a result, there is a great deal of competition among scientists for access to the facility. Often, in order to gain access, it is necessary to write a detailed advance proposal about what research is to be done, and what the scientists hope to learn from it. Originally, scientists had to convince the public that funds should be appropriated for the building of the facility. But once it is built, access is controlled by other scientists, who must be convinced by scientific rather than social arguments.

In this competition for access, it often happens that several groups of scientists make similar proposals because there are usually only a few experiments that need to be done at a particular stage in any science. Those who decide about access often unite several groups, both to avoid shutting out one group and to maximize the funds available. Alternatively, a group of scientists that is not confident of gaining access may join with a stronger group, trading a small chance for individual achievement for a larger chance at a part of the group's achievement. Often these strategies lead to the formation of needlessly large groups of collaborating scientists. Recently a committee of European physicists had to decide who would have access to the new particle accelerator being built in Switzerland. The decision resulted in the formation of groups containing hundreds of physicists from many countries who will collaborate to use the accelerator.

In areas of science other than particle physics and astronomy, very large experimental groups are much more rare. Nevertheless, there has been a substantial shift to larger research groups in almost all fields. The groups usually include one or more senior scientists, several junior scientists who have recently obtained doctorates, and a number of graduate students. Technicians may be involved as well. Often the output of the group is published jointly as the product of the laboratory, and it is sometimes difficult to know who is actually responsible for an individual piece of research. The image of the scientist as a solitary worker seeking truth in a lonely laboratory has become obsolete in experimental science.

Collaboration can be an extremely rewarding experience. The need to coherently explain one's ideas to an interested co-worker helps sharpen the thoughts. Co-workers can also give one another a psychological boost at critical times during the course of solving a difficult problem. In my own career such psychological reinforcement has been essential on many occasions.

Many other areas of science will take the same direction as experimental particle physics and carry out complex experiments in groups. If new and dramatically more expensive tools become available to biologists, for example, they too will carry out their research at large national and international laboratories as physicists and astronomers do.

Whether physics will continue to move in the direction of ever larger research groups working at ever larger laboratories depends on future discoveries and what they will tell us about the world. Particle physicists are constructing accelerators which, together with their associated detectors, are to be staffed by groups containing hundreds of physicists from many countries. As the cost of accelerators increases, the number that can be built decreases—as does the number of independent groups that can gain access to an accelerator. This development could sharply limit the number of novel experiments that can be tried. There are already proposals for a "world" accelerator, which would be shared by physicists from all countries. Most likely financial constraints, if nothing else, will limit the size of accelerators; those currently being planned are estimated to cost several billion dollars each.

Of course, the picture would change considerably if the cost and scale of accelerators and particle detectors could be drastically reduced without compromising performance. There are possibilities for doing this; one such is the use of intense laser beams to accelerate particles. If accelerator costs did decrease drastically, then each research laboratory could have its own accelerator, much as was the pattern with cyclotrons thirty years ago. If this happens, future research in particle physics might be done by small teams.

Another discovery that might break up the large groups of physicists would be the detection of fundamental phenomena beyond the research of accelerator physics. Particle physicists may shift their attention away from accelerator physics if some fossil remnants of the early universe too massive to be produced by accelerators are found. While some aspects of the study of such fossil particles might require large research groups, other aspects will require novel experimental approaches, and these might best be done by small groups. If the focus of particle physics shifts in this direction, the great laboratories of today may become like the great pyramids, once the center of attention but now just a stop on a guided tour of past glories.

A glut of scientists also leads to a competition for ideas. When there were few scientists at work in any field, it was possible for an individual to stake out one part of the field as his or her own, and to explore it without much competition from other scientists. This is not to say that there were not times when several scientists were working on the same problem at once. Several important discoveries have resulted from such competition, such as the discovery of electrons in 1897, but before the twentieth century, and even before the post–World War II period, it was rare.

Today the number of researchers in most areas of science is so large that any good idea is soon seized upon by many. This happens more readily in theoretical science, where there is no serious problem of access to facilities, but it also occurs in experimental science, through the duplication of similar research facilities at different laboratories around the world.

Competition often quickly pinpoints the strengths and weaknesses of new ideas, and helps avoid situations where a scientist might follow a blind alley for a long time. Many contemporary scientists thrive on this type of competition. In his book *The Double Helix*, James Watson tells how the structure of DNA was unraveled in a spirit of brisk and energizing rivalry. Competition can also reassure scientists that what they are doing has some value in the overall scheme of science, a reassurance not always available to the lone workers of the past.

There is also a negative side to increased competition. Often it leads scientists to work with extreme haste, which can lead to errors. Also, if scientists believe that there is less time available to complete their work, they will tend to produce partial results rather than a single complete picture, and important aspects of the phenomena being studied may get lost. One indication of this is the proliferation in several fields of science of "letter journals"—periodicals that publish brief communications very rapidly. In certain fields where competition is intense, the standard practice is to submit a series of brief progress reports to these journals, as a way of keeping ahead of the competition. Eventually the group is supposed to submit a complete description of the work to a journal of record, but this does not always happen, and when it does it is often too late to influence developments in the field. To this extent, the proliferation of these journals has led inadvertently to a decrease in the quality of scientific research and its dissemination.

Often outsiders complain that competition makes scientists afraid to reveal their results prematurely to possible competitors. Such secrecy is hardly new. In the seventeenth century, scientists such as Galileo submitted their work to colleagues in the form of ciphers, presumably in order to avoid having it stolen [Figure 30]—and some of the brief communications favored by today's scientists have this same cryptic quality. This secrecy is the antithesis of the relationship that scientists are ideally supposed to have with one another, and it is less common in science than outsiders would expect. Although a given scientist may be wary of describing his

A.   HAEC IMMATURA A ME IAM FRUSTRA LEGUNTUROY
(These are at present too young to be read by me)

B.   CYNTHIAE FIGURAS AEMULATUR MATER AMORUM
(Venus emulates the shapes of the Moon)

C.   MACULA RUFA IN IOVE EST GYRATOR MATHEM, ETC.
(There is a red spot in Jupiter which rotates mathematically)

FIGURE 30   Galileo Anagrams. A Latin anagram shown in *A*, was sent by Galileo to Kepler to "announce" the discovery of the phases of Venus (the literal English translation of the sentence lies below it). In *B*, the rearranged message, together with its English translation is given. In *C*, an "incorrect" solution of the anagram found by Kepler is reproduced, together with its translation. Remarkably, there is a rotating red spot on Jupiter, discovered 250 years later.

or her tentative results to a competitor, most scientists speak quite candidly about what they are doing. Because of this candor, and because of the active communication links that exist among most scientific centers around the world, it rarely happens that an interested scientist remains unaware of what other scientists in the field are doing. A more serious problem is informational overload. Each scientist is flooded by so much information that wading through it all requires immense expenditures of time and effort.

A more malignant situation, when it occurs, is scientific fraud. Recently, there have been a few prominent cases in which scientists have fabricated data; one such was the painting of patches on mice to demonstrate that grafts had successfully taken. It is tempting to attribute such pathologies of research to an increased competition for funds or a lust for prestige, but that is not usually the case. The culprit is most frequently self-delusion. Scientists become prematurely convinced of their ideas, and so take unacceptable shortcuts to demonstrate what they believe to be correct. These self-delusions are not new in science, and there is no evidence that their incidence is on the rise. In fact, the increased number of scientists working on any problem often leads to a quicker uncovering of such deceptions when they do occur.

Ultimately, there are only two defenses against deception in science. One is that the members of the scientific establishment must continue to remind themselves and their students of the acceptable practices in their discipline. On the whole, this instruction has been very successful in avoiding deceptive practices. The other defense against fraud is the principle that all scientific results must be held in doubt until they have been checked by several independent scientists—no matter how sterling the reputation of their proponents. This principle, which a healthy spirit of competitiveness can reinforce, ensures that almost all scientific scams will be quickly exposed.

The Funding of Science

The working lives of scientists have changed considerably because of a great increase in the funds available to support scientific research. These increased funds have made it possible for scientists to plan and carry out research projects that would otherwise be unimaginable. In experimental particle physics, some individual experiments have cost tens of millions of dollars, more than was being spent on all of science fifty years ago. Astronomical facilities such as the orbiting space telescope, costing hundreds of millions of dollars, are being built. In chemistry and biology, individual projects of this scope are uncommon, but the overall level of expenditure per scientist working in the field is similar to that in physics. In spite of complaints by scientists about the stinginess of government support for basic research, the support has resulted in a radical transformation in how science is done. Without government funding, much of what we know about the world would not have been discovered.

Government largesse has made the careers of scientists and the ambiance of research laboratories more pleasant than they would otherwise be. Yet, although the funds devoted to scientific research are large by previous standards, they are still limited. By an extension of Parkinson's Law, proposals for scientific projects expand until they go beyond the available funding. Furthermore, like a small population suddenly exposed to a rich new food supply, scientists have responded to increased funding by reproducing themselves rapidly. It has been estimated that the "generation time" for scientists—i.e., the time it takes one scientist to train a new one in the same field—is less than ten years, and scientists working at universities "reproduce" themselves many times through the students they have trained. The number of doctorates awarded each year is substantially larger than the number of scientists who die or retire; the scientific population has not yet reached zero population growth.

The many scientists involved in basic research, most of whom are dependent on government funds for their research, compete intensely for the limited funds available. In the United States, these funds are usually awarded through a process called "peer review." A scientist who wishes to have his or her research supported by the government must submit a proposal to the proper government agency, such as the National Science Foundation. This proposal is reviewed by a group of scientists working in the same general field, and compared to proposals vying for the same funds. Eventually, a number of the proposals are funded and the rest of the scientists are politely told to try again next year.

Because funding is so crucial, scientists spend a good deal of time and effort preparing their proposals. Often, as insurance against rejection, a scientist submits similar proposals to several granting agencies. Some scientists estimate that they spend as much as 20 percent of their working time preparing proposals. The "food supply" of government funds for scientific research is becoming relatively scarce and scientists must now forage for support, in competition with large numbers of their colleagues. This is not because of cutbacks in the available funds; rather, the immense increase in the number of working scientists and the extended scope of scientific projects have placed greater demands on the research-funds pie.

The pattern of future government support will be of great importance to the future of science. It is clear that we cannot expect any future increase in government support of science in a proportion comparable to the post–World War II increase. Nevertheless, there could be significant increases in the research funds governments provide. A substantial reduction in military expenditures, for example, could allow governments to appropriate more funds for scientific research, especially in the developed countries. In these countries, many technically trained people work on military-related problems, and their talents could be redirected toward pure research. Even without a change in priorities about how government funds are spent, continued growth in the real income of the people of the world may allow substantial increase in governmental support of pure research.

While it is likely that the level of government support of science will fluctuate and that valuable scientific projects will therefore be postponed or go unfunded, it is unlikely that there will be any major decrease in the level of government funding in the developed countries. Governments have recognized that scientific research is an important element in the economic and technological well-being of any country. It is no longer possible, as it was in the nineteenth-century, for the untrained basement inventor to carry the burden of advancing technology.

I can think of no scientific projects whose delay by a few years would seriously compromise the overall progress of science. Objective observers should be skeptical about predictions of impending doom from scientists whose projects are delayed for lack of funding. On the other hand, there could be lasting damage to science and society if a whole area of research were denied funding for several decades. Most scientists who wished to work in such a field would be forced to abandon it—and there might not be enough scientists left to carry on should funds become available again. Scientists in the United States who wish to use space probes to study other planets face this threat today. A combination of high costs, long lead times, and ineffective justification of scientific proposals have virtually halted the United States' efforts in this field. Teams of scientists, both within NASA and outside it, may soon break up, and planetary exploration by space probes could be stalled until the twenty-first century.

Yet there is one factor that will help diminish such prospects. The political diversity of the world is great enough that most worthy scientific projects are likely to attract financial support somewhere. In the 1970s, when the United States cut back on its space sciences program, several other countries took up the challenge, realizing that this was their chance to take a leading role in space exploration. Western European countries have a joint program to probe Halley's comet during its 1985–86 appearance, a mission that the United States canceled. This type of role exchange between the scientific enterprises of various nations is healthy for science.

Scientists must learn to look for alternatives when government funding diminishes or becomes uncertain. Molecular biologists are cashing in on the immense technological applications of their work, obtaining funding that is independent of government grants. Some of it comes from existing corporations involved in medical technology. Also, some groups of scientists have formed their own companies that will both carry out research and market the technological results.

It is possible to do first-class research in science as part of a profit-making enterprise, as demonstrated by the Bell Laboratories in the United States and Philips Laboratories in the Netherlands. However, there are also potential problems. If frontier research in biology is supported in large part by private sources, secrecy may come to play a much greater role in biological research than it ever has in physics research. The aim of maximizing corporate profits may conflict with scientists' aim of communicating with one another. The history of Bell Laboratories shows that this need not be the case. Nevertheless, if the profit motive does lead to substantial restrictions on communication, it would represent a real threat to the progress of biology. No research laboratory, no matter how large and distinguished its staff, can maintain a high level of research without cross-fertilization from the outside world.

Scientific Communication

At present, the most common way that scientists communicate with one another is through articles in scientific journals. Every reputable journal reviews each contribution for scientific soundness, rejecting any article it considers to be wanting. Also, the journals are readily available to any interested scientist. The enormous growth in the volume of sound scientific research has made it difficult or impossible to keep up with all that is published, even within a narrow specialty.

A problem particular to rapidly changing fields is that findings are often published too late to be of much use. Fifty years ago it didn't much matter if one learned a year later what other scientists were doing, but in today's highly competitive environment, a time lag is a serious disadvantage. Scientists must keep abreast of the latest findings. A long lag in publication can also increase the possibility of independent duplication of research—and a resulting priority dispute.

One response to this problem in some fields has been a rise in the distribution of "preprints," copies of articles distributed in advance of publication. Some scientists distribute hundreds or even thousands of copies of a single preprint. It is not uncommon for prominent scientists working in active fields to receive several preprints daily. The device of the preprint solves the problem of spreading information rapidly, but it does not necessarily do so equitably. It can be difficult for scientists who are not at major research centers to get access to all of the preprints that might be of interest to them.

Another common method that scientists use to exchange results is the scientific lecture. Lectures often describe developments sooner than publications do, because it is more permissible to make conjectures and describe less complete results. Presenting a lecture also gives a scientist early feedback about his or her work, perhaps even helping the scientist to avoid the embarrassment that would arise from publishing conclusions that turn out to be erroneous. Yet such lectures, like preprints, are usually available only to those scientists working at major research centers, since the lectures often serve the purpose of advertising a scientist's work to those he or she believes can render some professional service.

Groups of scientists often meet for several days, both to hear lectures by experts in a particular field and to share their progress in an informal setting. When these meetings are small, they are generally regarded as one of the best methods for scientific communication. But an effort to keep such meetings open to all scientists, combined with the increase in the number of working scientists in almost all fields, has made it difficult to maintain the intimate gatherings of even a short time ago. A biennial conference in high-energy physics, which began thirty years ago at the University of Rochester as a meeting of fifty people, now has become a major international event with thousands of participants. Various institutions and even nations actually vie for the right to organize it. Such megameetings have their own cachet, but they are much less likely to be a useful means for communication between scientists.

One response to this situation has been the subdividing of science into narrower specialties. Scientists interested in some small area have their own meetings and publish their own specialized journals. While this can help solve the problems of scientific overpopulation, it has the disadvantage of creating artificial walls between fields of science that may actually have important bearing on one another. Other solutions need to be developed.

The Universal Computer Network

A universal computer network will be a powerful new means of communication among scientists. Ideally, each scientist would have a computer terminal that would give him or her access to the work of a large number of other scientists working in the same field. To communicate a finding, the scientist would send it out over the computer network, where it could be retrieved by anyone interested. This would solve the problems of those not on preprint mailing lists or conference invitation lists. It would be easy to index and abstract each contribution, so that network members would be able to decide which contributions might interest them sufficiently. The network would be interactive, so that scientists could "converse" about each other's contributions via computer, thus giving them the type of feedback that takes place at conferences. An effective computer network would increase the pool of those familiar with the latest developments in any field, and the brainpower available to concentrate on outstanding problems would increase considerably.

Setting up and maintaining such a network would be expensive, but the seeds of such networks can already be seen in some areas of science, such as particle physics and nucleic acid sequencing. A successful network among designers of computer microchips operated for several years in the 1970s. Many types of computer networks will be in widespread use by the year 2000. The present cost of disseminating scientific information among scientists is already very large. In my department at Columbia University, the yearly cost for each scientist to print and distribute his work is about $1,000. For all the physicists in the world the figure is at least $50 million each year. Exchanging scientific information through a worldwide computer network might not cost much more, especially if the network is already in place for other uses.

Yet such a system would not be without its pitfalls. It might reinforce a kind of faddism that already exists in some areas of science. Sometimes, a large number of scientists will follow a specific line of research for a while, then quickly desert it in favor of something else. For example, some years ago, the discovery of a new form of water, dubbed "polywater," was reported by one scientist. Many other scientists began to study polywater, but after a while, polywater turned out to be just a contaminated form of ordinary water, and the research was quickly abandoned. In some cases, this type of intellectual migration leads to important progress. In other cases, it is more a matter of following the latest intellectual fashion. Preprints have contributed strongly to such fads, since they allow a rapid dissemination of scientific works-in-progress that can inordinately influence impressionable scientists. A computer network would magnify this process.

Great self-confidence is required for scientists to disregard the line of research approved by their colleagues or by the intellectual leaders of their fields in favor of pursuing their own ideas. Such scientists run the risk of having their work largely ignored. At present, scientists can at least publish their work and hope that a shift in fashion will occur. If all communication were via the computer network even this hope would disappear—because of the more rapid feedback, or lack of it—and the pressure to conform to the favored topics of research would increase.

Another problem that would be aggravated by a universal computer network is the increasing demand on the individual scientist's time. The ease of conversations via the network would actually decrease the time left over for research. It would be as if each scientist were attending a perpetual conference with colleagues everywhere. A scientist would have to exhibit great self-discipline to limit the use of the network in favor of his or her own research.

In spite of these possible flaws in its operation, I expect that computer networks will become the dominant means of communication in the twenty-first century.

The Place of Science in the World

The role that science plays in the overall intellectual life of society has changed significantly over the years. At the dawn of modern science, in the seventeenth century, it was common, even for scientists like Newton, to think of science as subordinate to other intellectual pursuits, such as religion. Scientists now consider that scientific inquiry dominates the human intellectual adventure. This change of view is largely a response to the unparalleled success that scientists have had in solving the problems that they pose. The contrast between science, where the answers to questions are often found quite rapidly, and fields such as philosophy, where definitive answers are almost never found, has acted as a powerful influence in convincing scientists of the intellectual preeminence of their work.

This supremacy meets with very little challenge from nonscientists. When something new or unexpected happens in the world, such as a summer cold spell or a new disease, the immediate reaction among many is to ask what science has to say about it. This attitude is not really based on an understanding of what a scientific explanation is, or how scientists go about trying to understand new phenomena. Instead, it comes from some general aura of prestige that is based vaguely on the past achievements of science.

Yet there are exceptions to the perceived intellectual predominance of science, especially in the areas of ethnics and religion. Many scientists believe that science has no way to answer ethical questions, such as whether abortion is morally justifiable. All science can hope to do is help clarify some of the factual matters related to these questions, such as whether a fetus can feel pain. Many nonscientists and some scientists consider it a fault of science that ethical matters are beyond its grasp; after all, ethical questions are at least as important to the lives of people as are factual ones. This type of criticism goes back at least as far as Socrates. Yet many philosophers argue, as do I, that ethical questions do not have answers in the sense that scientific questions do; scientists show good sense in not using science to try to answer them.

The relation between science and religion is more complex, and the final lines between the two have not yet been drawn to everyone's satisfaction. Many of the topics that were once matters of religious faith have been transferred to the purview of science. This change has not happened without bitter arguments, over, for example, the evolution of species, molecular biology, and in some places, over the whole outlook of science.

It has been demonstrated over a long history of conflict that religion has nothing to teach science about the matter that most concerns science, that is, the way the universe works. Accordingly, it would be best if religious believers abandoned their efforts to influence the content of science. The long-term trend is certainly in that direction, and despite a flurry of legal activity among those calling themselves creation scientists, it is not from religious believers that the most serious challenges to science will come in the future.

It is more common for religious believers to preach to scientists about the moral aspects of science, particularly when its technological applications are as far-reaching in their impact as some forms of biotechnology, which could literally change the human species. Everyone should bring whatever sources of moral inspiration they have to bear on such critical questions. But even here, religion has no more claim on the attention of nonbelievers than any other source.

If religion has nothing useful to say to science, does it follow that science has nothing useful to say to religion? Much early religious belief offered explanations, such as those in the book of Genesis, of aspects of the universe that are now explained by science. To the extent that this motive still plays an important role in religious belief, the results of science are relevant, and when understood by the believers, could lead to a modification or abandonment of the belief. Scientists in non-Marxist countries rarely try to systematically explain the consequences of their work for religious belief, and I do not expect this to change.

The "Image" of the Scientist

On the whole, nonscientists in developed countries have had positive attitudes toward scientists, at least in the past fifty years or so. Opinion surveys among adults in the United States tend to give scientists very high ratings as desirable contributors to society. This endorsement derives mostly from the technological applications of scientific research, rather than from any real understanding of what scientists do. In fact, a strong case can be made that the public at large has little sympathy for what motivates scientists in their work. As long as there is a continued appreciation of the technological by-products of this work, this is not a serious situation. But continued appreciation is by no means assured, especially since some technological developments—nuclear reactors, means for prolonging life, and computers, for example—have been denounced by many.

Even if we can rely on public acceptance of science for its technological by-products, another problem must be faced. If science is to continue as a living activity in the future, it will need new scientists to replace those now functioning. Although we are presently training more than enough future scientists, this may not continue. Young people decide to devote their careers to science for numerous complex reasons. Certainly included among these reasons is exposure at an early age to the successes of science, presented in such a way as to make it seem both desirable and possible to contribute in a similar way.

To some extent, such an exposure to science is given youngsters in school and in books. However, an alternative, and largely negative picture of scientists has been presented for a long time to most children, at least in the United States. In comic books, movies, and television, the dominant portrayal of scientists to children is highly unsympathetic. For example, the description of

scientists given in television cartoons bears no resemblance to anything in the real life of scientists, and the differences all make the
motives of scientists seem worse than they are. The "mad scientist" of these television cartoons, who is trying to rule or destroy
the universe, is but one manifestation of the negative image of science presented in the mass media. The television series *Star Trek*,
has a wide following, especially among the young. It describes the
"voyages of the starship *Enterprise*, her five-year mission to explore strange new worlds, to seek out new life and new civiliza-
tions, to boldly go where no man has gone before." This sounds
like a scientific enterprise of the first order, and its followers may
take the plots as an indication of what scientists do and think.
However, almost all of the actual plots in the series involve violent
conflicts, often between the crew of the *Enterprise* and various
"scientists" with very nonscientific motivations. The only vaguely
scientific character presented at all sympathetically is Mr. Spock,
who is only partly human.

Even when the motives of scientists are not being misrepresented, their life and work is generally described, in mass media
fiction, in such negative terms as to make science seem a highly
undesirable career. Not surprisingly, opinion polls show that most
young people end up with this very conclusion about science.

It is interesting to speculate about why the purveyors of popular culture should make such strenuous efforts to paint science
and scientists in the lurid tones that they usually do. It is not clear
whether the negative picture of science is drawn purposefully to
downgrade science, or whether it is done simply as a convention of
the media. This pattern is probably related to the widespread mis-
understanding in our society of what scientists are trying to do.
Whereas most scientists are concerned with understanding the
world, an activity which should appear benign even to the pro-
ducers of television cartoons, many people unfamiliar with science
assume that scientists are mainly concerned with controlling the
world. It is a short step to the conclusion that scientists are motivated to rule the universe, the primary motive attributed to them
in the popular culture.

It is unlikely that this image of scientists will change anytime
soon as a result of increased understanding of what science is
about. There has been some increase in scientific literacy in the
general population, as witnessed by the success of several attempts
to "do" science in the mass media, such as "Cosmos" on television

and the widely distributed magazine, *Discover*. However, these have largely been directed to adults, rather than to the children who are the victims of media misportrayals of scientists. Scientists do not take this problem of the misrepresentation of science seriously, but we are mistaken in our complacency. Whenever a false image of one group is systematically presented to impressionable minds, there are bound to be long-term consequences to the disadvantage of the misrepresented group. If this has not yet happened to scientists, it is because of favorable pressures coming from other directions, but we cannot rely on such offsetting factors to always prevail. If scientists do not pay attention to what others are falsely saying about us, we may soon find that the consequences are very unpleasant.

Internal Threats to Science

Whether science will even survive depends on factors both inside and outside science. I believe that the internal factors are the critical ones. Science has flourished in the past without significant external support, and could probably do so again if necessary, though on a very reduced scale. But it cannot flourish in the absence of certain attitudes among scientists themselves.

Different human impulses have contributed to science, in the form that it has taken since the seventeenth century. One is the desire to understand the world through the use of our intellect. This impulse is found not only in science but in other human activities, such as speculative philosophy and some forms of theology. A second impulse, which is more specific to science, is the unwillingness to accept results solely on the basis of authority. Any scientific idea is fair game for investigation by each person who enters into science; the fact that a particular idea was invented by a revered scientist of the past does not make it true. In this regard, science is different from theology and certain other intellectual systems, where the authority of the founder is often taken as sufficient reason for the validity of a proposition. It is the experimental character of much of science that is the purest expression of this aspect of the scientific attitude.

Paradoxically, the third impulse that has entered into the making of science is a willingness for the scientists of each generation to build on the work of past scientists. This cumulative character of science distinguishes it from speculative philosophy, where each philosopher who considers a problem tends to regard the work of predecessors as underbrush to be cleared rather than a solid basis for his or her own investigations. The fact that science is cumulative is one aspect of the collaborative spirit of scientific research; another is that the work of each scientist depends on that of contemporary scientists, both through their inputs to his or her own work and for its ultimate acceptance by the community of scientists.

Arguments have been given, for example by Gunther Stent, that the impulse of curiosity that has motivated much scientific research is waning, along with other human impulses toward creative achievement, and that because of this the future world will be one in which scientific understanding is no longer a matter of concern to anyone. It has even been suggested by the American physicist Leon Lederman that future societies will discover new pleasures so intense that none of the activities that we presently value will be able to compete with them for our attention. This suggestion grows out of the discovery that under some conditions, rats will choose electrical stimulation of the pleasure centers in their brains over the traditional sources of rat gratification, such as sex and food. The notion that the ultimate function of conscious mind on earth is to devise means to turn itself off is one that I find ludicrous.

Furthermore, a society composed of pleasure-seekers would be at a severe disadvantage in competition with others whose members were differently motivated, a point realized too late by the inhabitants of ancient Sybaris. On the whole, I think that the falling off of human curiosity is the least of the threats to the continuation of science in the future.

Of course, what is considered a fit subject matter for curiosity can change. It happened both in ancient Greece and in ancient China that the main focus of interest among intellectuals changed from the universe outside to a study of human affairs and to the words produced by the human mind. Some have urged that a similar change of interest would benefit our own society. However, I would be surprised if this happened, unless there was an immense advance in the intellectual content of political and social thought. One of the attractions of modern science is the obvious possibility for incontrovertible progress within it, as distinct from the doubtful progress to be found in other spheres of thought. Socrates was able to divert Greek philosophy from science to morals because neither had made much progress in his time. It will take more than words to do the same in the face of the intellectual accomplishments of modern science.

I am also reasonably optimistic that future scientists can resist the temptation to substitute authority for observation and experiment as the ultimate source of scientific wisdom. In society at large, there are trends for and against submission to authority, and this diversity will probably ensure that many scientists will continue to follow their own paths to truth.

I am less certain that science will be able to maintain its sense of community, both with past science and among colleagues. One factor working to distort this sense of community is the increased competition for the resources necessary to do scientific research. When scientists were all relative paupers, such competition was subordinated to a sense of common purpose. The manna of government support, the rapid growth that this has made possible, and the subsequent absorption of the available funds has made competition a much more important factor in science than it once was.

Such competition is not altogether a bad thing; many scientists (like people in other fields) work best under the pressure it fosters. But we should not imagine that the sense of shared purpose that plays such an important role in the way science has been done in the past will automatically survive the transition to a social setting in which colleagues are regarded as rivals for essential resources. The sense of community with past scientists involves other issues, which I discuss in the concluding chapter.

External Threats to Science

The threats that science faces from outside itself are less serious than the internal ones discussed above, although they are more apparent to the casual observer. There is the threat of social control over the type of research to be carried out as well as over the conclusions of research. There is the threat of severe cutbacks in the level of financial support. Both of these threats have become real in some recent cases, and the experience has conditioned many scientists to a state of alertness about their recurrence.

I believe that the only serious threat from society to the continuation of science comes from the will to exercise social controls over science. This will is to be found in various places in society. It exists among some of those who feel threatened by the discoveries of science, such as the believers in fundamentalistic religion. It exists among those whose passion for social justice blinds them to the possibility that there are other human impulses whose gratification is equally desirable. It exists among those who do not believe that some pursuits are justified in their own terms rather than as means to some other social ends. And most seriously, it exists among those who are fearful of the possible consquences of trying to use applied scientific knowledge to modify the human condition. These are very distinct rationales for limiting the activities that scientists can pursue, and the degrees to which they can be justified vary significantly. But for the purpose of predicting future restrictions on science, what matters more is the strength of the impulse, and how it may manifest itself.

I believe that there is no valid reason for any society to limit the type of questions that scientists may investigate, or to constrain the type of answers that scientists may find to the questions that interest them. Society should only restrict scientific research that would directly harm other human beings. A society is no more justified in regulating the curiosity of its scientists than it would be in regulating their eating habits, or regulating the expression of its artists.

A serious problem for science today is posed by the actions of those people both inside and outside science who wish to forbid research in certain controversial areas. Scientists working on the inheritance of intelligence, and on recombinant DNA have met with sometimes violent opposition. As a result, many interested scientists have decided to avoid these fields altogether. One prominent researcher on the heritability of intelligence has said that if her work showed that intelligence was largely a matter of heredity, she would be tempted to leave the United States because of the controversy that this result would generate.

The aims of science are never served by this type of coercion. Though there may be valid reasons for trying to persuade scientists of the inappropriateness of certain lines of research, and though sharp intellectual criticism is well within the bounds of acceptable relations among scientists, violent harassment of fellow scientists of whose research one disapproves lies far outside those bounds. Scientists as a group should take steps to discourage such coercion when it arises inside of science. Those scientists who regularly engage in it should be regarded with the same contempt as those scientists who falsify research. Whatever their other merits, they should be penalized professionally by their colleagues. If we allow anyone, whatever their motives, to repress scientific curiosity by force, then one of the great lessons that science has taught mankind will have been lost: Unless we are all free to express our curiosity to seek the truth, wherever it leads us, then none of us are truly free.

One area in which I believe that society does have a legitimate concern is over the technological consequences of scientific research, but this concern is not a justification for forbidding a certain type of research. We cannot know very well what new technology can emerge from scientific research until after the research has been done, often not until long after, so that trying to control technology by controlling science is likely to be ineffective, unless we restrict all research. It is much more effective to control the technology itself, after its possibility has been demonstrated, but before it has been developed or implemented on a large scale.

A different question is whether society should influence the direction of scientific research, in accordance with criteria lying outside of science, by preferential funding of certain areas of research. In this model, scientists would be free to investigate what areas they chose, but they would be more likely to get funded if they worked on certain subjects valued by society. To some extent, this model actually operates, at least in the United States, under the name of targeted research, which aims to achieve specific technological goals. But most scientific research is not presently targeted, and scientists who wish to follow their own interests can usually still obtain funds for their research by convincing their peers that what they are doing is worthwhile according to scientific criteria, rather than according to externally imposed criteria.

This situation, which I regard as very desirable for the progress of science, may not last. When faced with the present situation in which there are increasing numbers of scientists competing for funds that are not increasing equally rapidly, it is difficult for any society to resist the temptation to choose to support those projects that are perceived as most likely to bring benefits in the short run to most members of the society.

It would be highly imprudent for society to attempt to target most research toward specific technological ends, however, because we cannot accurately predict very far in advance which approaches to unsolved scientific problems will lead to the desired technological results. Evidence for this point of view comes from comparing the Apollo Project, which required no new scientific discoveries, and which was completed in about the estimated time, with the "war on cancer," which requires new discoveries, and which has not yet been successful. Even today it is not clear what the best approach to finding a cancer cure should be. Unfortunately, scientists themselves often promise that specific technologies will come out of their basic research, presumably as a way of encouraging funding. Such promises, unless they are meant as generalizations about the ultimate consequences of research, or refer to very specific situations that might more aptly be labeled "development" than "research," are short-sighted. In the long run they do not serve the aims of increasing support of science because these predictions of technological success are often wrong, and when this happens, it can undermine public belief in the accomplishments of science—and the truthfulness of scientists.

When new scientific discoveries are needed in order to achieve some particular technology, we know of no better way of making these discoveries than through the process of ordinary, untargeted research. Sometimes these discoveries will come in areas clearly related to the desired technology, sometimes in seemingly unrelated areas. This attitude may appear to be self-serving on the part of scientists who wish to follow their own interests. However, it is the best we can do, and is the approach that I believe will ultimately prove most fruitful.

The Future of Science

The history of science has shown many broad changes, both in the scope of individual sciences, and in the overall character of science. It is logical to expect further changes. Some of these changes will be continuations of past trends. But there are other, more radical developments coming, including changes in the very notion of what science can explain.

The coming developments in physics and biology that I have described will, in part, determine the overall scope of these sciences. In the case of physics, there will be a continuation of the trend that has increased the range of phenomena with which physicists deal. After physicists discovered the laws that govern familiar phenomena and familiar forms of matter, their attention turned to the unfamiliar, to subatomic particles with fleeting lifetimes, to the earliest stages of the universe, and to distances beyond the range of our strongest telescopes. Miraculously, they have found it possible to encompass many of these unfamiliar phenomena into laws that are outgrowths of those we already know.

This pattern in physics will accelerate in the future, with a continued focus on the phenomena out of the range of ordinary experience. In the short run, this can probably be done incrementally, without too much of a change in the basic ideas of physics. Eventually, as our reach extends ever further beyond the familiar, we will need to modify basic ideas in a fundamental, and presently unimagined way.

For biology in the coming period, the main work is to find answers to many of the open questions about living things. I expect this to be largely accomplished through the systematic extension of molecular models, through the use of novel experimental tools, and most importantly, through the introduction of appropriate symbolic (i.e., mathematical) descriptions of biological phenomena. Biology will become more like the physical sciences, reasoning from general principles to specific cases. The use of sophisticated methods for data processing will also play a greater role in future biology, and some of its most important future insights should come from the new information that this extra data will provide.

It is also possible that future biology will deal with an increased range of phenomena when and if we discover extraterrestrial life forms. But it is also possible that novel forms of life will be discovered on earth or even created in our laboratories. Any of these developments would have positive effects on biological theory, as they would provide new tests for its fundamental ideas as well as new perspectives on what is already known.

Finally, the active field of biotechnology will have a great effect on future biology. The development of useful products will be an important new criterion for how well biologists understand the phenomena they study. This quest for concrete applications will force biologists to be much more attentive to the details of their explanations, just as physical technology has done for physics. It is also likely that some forms of biotechnology will provide new experimental tools for biologists to use in their studies.

Changes in What Scientists Accept as an Explanation

If it were possible to bring to life some of the illustrious scientists of the past and describe to them the present state of science, they would be impressed by the experimental discoveries that have been made since their time, and by the new means of observation that now exist. They would, I think, be less happy about the way that theoretical science has progressed—especially because of the profound changes in the types of explanations that scientists are willing to accept for the phenomena that they study. Even many contemporary scientists find some of these changes disquieting. But it is important to recognize that changes in the mode of scientific explanation—and associated dissatisfaction by some scientists—is nothing new.

From the time of Newton to the end of the nineteenth century, the mechanical model, involving small bodies moving through space under the influence of reciprocal forces, was considered the ideal form of explanation in physics. In the late nineteenth century, James Clerk Maxwell used mechanical models to help him find the equations that describe electromagnetic phenomena [Figure 31]. Other scientists of the time, such as Lord Kelvin, found the models more convincing than the equations Maxwell derived from them.

Ironically, Maxwell's successors found that they could do better by keeping the equations and abandoning the mechanical models. Some of these physicists replaced mechanical models by a description in which all phenomena were to be explained in terms of electromagnetic fields. This effort, which was never completely successful, came to be known, in the early twentieth century, as the electromagnetic world picture.

An even more radical shift came in the 1920s with the development of quantum mechanics. Here the assumption that physical phenomena take place independently of how they are observed was called into question. This basic assumption of all of earlier physical investigation was replaced with a description in which what is seen to happen depends critically on the means that one uses to look at it. Additionally, with the advent of quantum mechanics, belief in the possibility of an *exact* prediction of the future has generally disappeared.

In the face of such radical changes in what a physical theory can do, and in the objectivity of our description, it is not surprising that there has been strong resistance among some physicists to quantum theory. Einstein, who himself made important contributions to the early development of quantum theory, was the leader of the loyal opposition to it, and some scientists today feel the same way. Their objections involve more than the vague sense that any scientific theory may someday be replaced. Instead, they argue that quantum theory does not satisfy their notion of what a theory should do.

It is highly implausible that the features of quantum theory that Einstein and others find objectionable will be eliminated in any future theory. When we confront unfamiliar phenomena, such as those on an atomic or subatomic scale, there is no reason to ex-

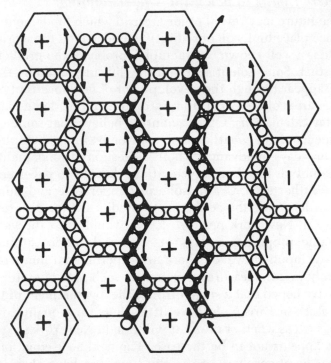

FIGURE 31 Maxwell's Model for Electromagnetism. The hexagons represent vortices, the small circles between them are particles associated with electricity. The particles and vortices are supposed to roll on one another and so produce the phenomenon that we describe as electromagnetism.

pect that we can maintain the notion that the properties of objects are independent of how we look at them.

It should not surprise us when we run into limitations on what we can know or predict. Although science aims to discover laws whose validity transcends their human origins, it is never possible to completely achieve this goal. Our brains—and what they can imagine—have evolved in order to deal with phenomena on a human scale, and we have an intuitive sense of what types of explanations are possible on that scale. But this sense can mislead us when we are dealing with unfamiliar phenomena on the scale of atomic or subatomic processes. Instead of following Einstein in his resistance to the indeterminacy of quantum theory, we should turn to one of his aphorisms, and marvel that any part of the world is comprehensible to us at all.

Are There Limits to Scientific Understanding?

The future may reveal other areas in which fundamental laws of science place limits on our ability to understand the world. One example is a well-known "intractable" problem, the prediction of the weather. Some scientists think the difficulty in doing this will soon disappear through the development of new means of observing large areas of the earth from satellites, and by doing elaborate computer calculations. Other scientists believe that some observable aspects of the weather may forever be unpredictable more than a few days in advance. Their suggestion of fundamental unpredictability is based on the equations thought to govern aspects of the weather such as atmospheric temperature and humidity.

In order to use the equations to predict the temperature and humidity in New York next week, we would need the values of those same properties today—at various locations around the world—as input. However, there are experimental limits on how accurately we can measure any quantity such as temperature. One might have hoped that a small error in the measurement of today's temperature in Paris wuld have little effect on a qualitative prediction, such as whether or not it will rain in New York next week, but this appears not to be the case. The nonlinear equations that describe the atmosphere are of the type that have chaotic solutions (discussed in Chapter 6), so that two minutely different situations for today's temperature in Paris, which cannot be distinguished by measurement, can lead to two very different patterns of weather in New York the following week. One scientist has described the situation as implying that next week's weather in New York may depend on the flapping of a butterfly's wings in Paris today! We do not yet know if this is definitely true, but a plausible case has been made for it.

Some of the phenomena currently being investigated by scientists, such as those involving the human nervous system, are so complicated that some scientists believe we will never be able to understand them in the sense that we understand a simpler system like a virus.

In evaluating such claims, we must distinguish between conclusions that are based on some general principle, such as Heisenberg's uncertainty principle, which explicitly restricts what we can do and know, and conclusions that are based on a vague sense that a specific problem is too complex for the human mind to comprehend. The idea that we cannot predict the weather far in advance because of the nature of the equations that govern the earth's atmosphere belongs in the first category. The suggestion that the human nervous system is beyond our capability to understand is, for the time being, in the second. The fact is, we do not yet know enough about the nervous system to be able to say whether we will eventually be able to understand it. To know that we could not, we would have to identify some aspects of the nervous system so complex that understanding them would necessarily resist any advances in experimental techniques and in mathematical analysis. This in itself would be a major discovery, comparable to that of Heisenberg's uncertainty principle.

Many scientists, including myself, deeply regret that future science may see more restrictions on what we are able to understand and predict. Most scientists enter into their studies with the expectation of extending human understanding, not of finding limits to it. On the whole, this optimism has been justified by the history of science. Yet, while we should not needlessly assume restrictions on what the human mind can achieve, we should remain aware of the possibility that the lack of predictability inherent in quantum mechanics may be only the first indication of a future science whose grasp will fall short of what scientists would have hoped.

The Rise of New Sciences

An exciting aspect of future science will be that new sciences will arise, encompassing areas of study not presently recognized as part of science. Such extensions of the domain of science can occur through the discovery of unsuspected phenomena so distinctive that they cannot be fitted into the categories of previously recognized science. An example was Leeuwenhoek's observation of bacteria through his microscope in the seventeenth century. Perhaps the discovery of extraterrestrial life forms will play a similar role in future science.

Another way in which new sciences arise is through the recognition that some aspects of several known sciences have enough in common to form a connected field of their own. This fusion can occur either to subfields of a single science or to aspects of sciences that have traditionally been considered different disciplines. For example, little progress had been made in understanding the phenomena of magnetism until Oersted's discovery that electric currents could produce magnetism. This led to the new science of electromagnetism, which helped solve outstanding problems in both subfields.

To identify candidates for future scientific fusions, we must look to areas of science where gaps in understanding are most apparent. If the scientific problems within a field cannot be attacked successfully by the methods used within that field, it may be that the phenomena being studied need to be understood within the framework of a wider array of phenomena, in a science yet to be invented.

One place to expect a new science to emerge is in the study of how order evolves from disorder. There are several places within present science where we can recognize that complex order emerges from simplicity—without knowing how this happens. The most interesting example is the origin of life, but other examples in physics and chemistry have been mentioned. These different examples of the development of order come about through distinct underlying mechanisms. However, there are relations among suggested mathematical descriptions of each case that suggest the possibility of finding a unified description of a whole class of such phenomena. Such a description would be analogous to the subject of thermodynamics, which applies to many different phenomena as varied as the emission of radiation by a star and the flow of ions through a cell membrane. Thermodynamics makes general statements that constrain what happens in all processes involving the flow of energy and matter. A general science of the evolution of order might make general statements about how order evolves under a variety of circumstances. These general laws might not completely determine the evolution of order in different circumstances, any more than the laws of thermodynamics determine how energy flows in every context. However, once such general laws become available, they could suggest the supplementary principles necessary to completely characterize individual cases.

A science of order could emerge by weaving together a number of threads. One is the study of the nonlinear differential equations, which may give criteria for the emergence of order in varying circumstances. Another is thermodynamics, which in some of its modern versions contains ideas about how a flow of energy impinging on a system from outside can lead to the evolution of the system, in the direction of increasing order. Still another thread may emerge from the experimental study of phenomena like the laser, which involve the evolution of order.

One ingredient of a new science of order that is missing is some general principle that expresses when a system will evolve in the direction of increased order. We know some of the requirements for this to happen, such as an external supply of energy. We also know of some conditions under which this evolution will take place. But no one has found a general way to identify which systems will become orderly and which will not. If such a principle can be formulated, it would suggest relations between systems made of different components, but which undergo parallel types of evolution.

A new science of order could help us find answers to many major scientific problems, such as how life originated on earth and how complex organisms develop from single cells. It could also suggest possibilities for complex order beyond those we have identified. These might include life forms that could thrive in environments inhospitable to the life we know. They might also include new types of inorganic order that we could construct, even if they do not occur naturally.

A new science of order will have major ramifications in science at large, and will change scientists' world view—perhaps as drastically as relativity and quantum mechanics have done.

The Demise of Old Sciences

Sciences can disappear as well as arise. A "science" can be found to be nonsensical, as happened to phrenology and astrology. More commonly, a science once thought to be autonomous may merge into another science. This happened in the seventeenth century to the astronomical study of planetary motions, which became a part of physics; it happened in the nineteenth century to optics, which became a part of electrodynamics; and in the twentieth century to much of inorganic chemistry, which became a part of atomic physics. In such mergers, the content of the subsumed science is not lost. Indeed, some of its outstanding problems may be solved through the merger.

When it becomes possible for the subject matter of one science to be studied more effectively through the application of the methods of another, scientists should regard this as progress. This will happen to all of chemistry in the near future. It is already clear that all of the phenomena of interest to chemists are manifestations of the electronic structure of atoms and molecules. Yet this does not mean that the subject matter of chemistry will disappear, or that chemists should lament. Instead, the scientists now calling themselves chemists will increasingly use the ideas and experimental techniques of physics to study the questions that concern them. Already, many chemists do research that is almost indistinguishable from what some physicists do. Eventually, chemistry and physics will be distinguishable not by their ideas and techniques, but only by the problems to which these are applied. At that point, there will be a single science of matter and energy, which is likely still to be called physics, but with many branches in which different aspects of matter and energy are studied.

Although biology may be subsumed into physics in the distant future, biologists need not defend themselves against the prospect that their science will soon merge into physics in the way that chemistry and physics will merge. Several aspects of the behavior of living things *have* been understood in terms of physical models, involving the properties of nucleic acid and protein molecules. This process of finding physical explanations for biological phenomena will continue. However, it is unlikely that we will soon be able to understand all of the activities of complex organisms just by applying the ideas of physics.

What will happen is that more and more of the ideas and techniques used by biologists will originate in chemistry and physics. If ideas that can be derived from physics are successful at accounting for the phenomena of interest to biologists, then biologists will adopt them. If ideas that biologists can generate without reference to physics are more successful, then future biologists will continue to use them. The ultimate criterion for the value of a scientific idea is not its origin, but its success in explaining phenomena under study.

There is a more troublesome way for parts of a science to disappear. When a new mode of explanation takes over a branch of science, as when molecular explanations of heredity came into use in genetics, there often remain unsolved problems that are disregarded. Sometimes it is realized that the old problem cannot be solved at all and that it is right to disregard it—as happened in physics to the question of what orbits an electron follows in an atom. But usually, a matter of scientific principle is not involved. Rather, scientists concentrate on the areas of their field that are most amenable to the new way of thinking. Occasionally, they return to "clean up" the older subjects, but more commonly, there is so much new material to be studied that the older unsolved problems fade away.

Science is an activity carried out by a finite number of people at any given time. These scientists can only focus their efforts on the problems of greatest interest to their community, which are often those generated by the latest discoveries. There is usually little incentive to study the application of new ideas to old problems, except when these problems are so well known or so strikingly important that they remain in the consciousness of all workers in the field.

This tendency has been exacerbated by the immense expansion of scientific information. As more comes to be known and new questions arise, scientists can learn an ever smaller part of what is known and can concern themselves with an ever narrower segment of it. One consequence has been that even when there has been no radical change of ideas, scientists concentrate on the most recently discovered parts of their field, and the burning questions of even the recent past tend to be forgotten. Science is thus in some danger of losing its character as a cumulative body of knowledge.

Although this tendency is hardly new, it has accelerated recently. The typical article or book cited by researchers is only a few years old. In my own field of theoretical particle physics, many references are to articles that have only been distributed as preprints. One reason for this telescoping of the focus of scientists' interests is that many problems are solved rapidly once they are properly posed. However, the desire of many scientists to work on the same problems as their fellows is probably a more significant factor.

There are trends that could act to mitigate the tendency to disregard past science. One is the immense increase in the capacity to store, analyze, and make accessible information through the use of computer data banks. When the universal computer network is in place, it will be possible to make available all of the scientific literature of the past. This will remove one barrier to using "old" science, but the problem of translating the concepts of the past into the scientific language of the present will remain. This might be a good project for computer translation, as it would seem easier for a computer to grasp the more precise vocabulary and syntax of science than that of ordinary language.

However, the most serious barrier to continuity of scientific interest is a lack of concern that the problem exists. Overcoming this is a matter of proper scientific education, and of the reward system of science. If young scientists are taught to be concerned about the unsolved questions of the past, and those who work on them are suitably rewarded when they succeed in answering one of them, then there will be continuity in scientific discovery. If this does not happen, then science will live more and more in the present, and the works of each generation of scientists will be unimportant to the next generation. This would be a sad ending to what has thus far been the most successful human effort to build an intellectual structure that spans the generations.

I conclude this survey of the future of science relatively optimistic about what this future will bring. There are many important unsolved problems in science, and there are good prospects that some of these will soon be solved. There are forces acting on science that will transform how science is done. New techniques for gaining information about nature are being developed that will lead to new discoveries as exciting and unexpected as any of the great discoveries of the past. Present and future discoveries in science can result in new physical and biological technologies that can transform human life. All of these represent the bright prospects of future science.

Further Reading

In this section, I describe some references that I have found useful in connection with the topics discussed in the text. The reader who wishes to learn more about some of these may wish to consult them.

Introduction

Page 14.   A report prepared by Vannevar Bush in 1945 proposed a number of steps for the government support of pure science. These proposals have guided much of the United States government's policy toward scientific research in the post World War II period. However, the report made no effort to predict the effect of the program that is proposed on the future of science. See *Science, The Endless Frontier* by Vannevar Bush; United States Government Printing Office, Washington, D.C., 1945.

Page 14.   The definition of science that I have given owes much to the ideas of Ernest Nagel, as described in Chapter 1 of *The Structure of Science* (Harcourt, Brace and World: New York, 1961).

Page 17.   The early history of radioactivity, and some of the questions it raised are summarized by Abraham Pais in his article, "Radioactivity, Two Early Puzzles," *Reviews of Modern Physics,* vol. 49, p. 925, 1977.

Page 18.   Some aspects of the history of the biological applications of X-ray crystallography are discussed by Horace F. Judson in Chapter 9 of *The Eighth Day of Creation* (Simon and Schuster: New York, 1979).

Page 18.   Eddington's fishnet analogy for scientific discovery is presented in Chapter 2 of his book *The Philosophy of Physical Science* (Macmillan: New York, 1939). The conclusions that he draws from it, in favor of a major role for a priori considerations in science, are not accepted by most other scientists.

Page 18.   The fascinating discovery of ecologies on the ocean bottom is described in an article by R. D. Bullard and J. I. Grassie in *National Geographic,* vol. 156, p. 689, 1979.

Page 19.   The possibility of other life forms on Earth, with widely different biochemistries, is discussed in a book by Gerald Feinberg and Robert Shapiro, *Life Beyond Earth* (William Morrow: New York, 1980).

Page 19.   A readable discussion of tensor analysis and its applications to physics is given in Chapter 25 of *The Evolution of Scientific Thought,* by A. d'Abro (Dover Publications: New York, 1950).

Page 19.   Catastrophe theory is discussed by its inventor, René Thom, in *Structural Stability and Morphogenesis* (W. A. Benjamin: Reading, Mass., 1975).

Page 21.   Newton's view is described in the General Scholium of Book II in his *Principia.* Einstein's is given in his lecture "On the Method of Theoretical Physics," reprinted in *Ideas and Opinions* (Crown Publishers: New York, 1954).

Page 22.   No detailed history of this era in particle physics has yet been written. Some aspects of this history are described in the Nobel Prize lectures for 1979 by Sheldon Glashow, Abdus Salam, and Steven Weinberg, reprinted in *Reviews of Modern Physics,* vol. 52, pp. 515–44, 1980.

Chapter 1

Page 31.   Some details about the subatomic particles, and other topics discussed in this chapter can be found in *The Cosmic Code* by Heinz Pagels (Simon and Schuster: New York, 1982) and *What Is the World Made Of?,* by Gerald Feinberg (Doubleday-Anchor Press: Garden City, 1977).

Page 34.   The expansion of the universe, and other topics in cosmology are discussed in Steven Weinberg's *The First Three Minutes* (Basic Books: New York, 1977).

Chapter 2

Page 52.   Some idea of the methods that biologists use can be seen in their original papers. A sample of important papers in the history of biology are reprinted in *Modern Biology,* edited by Elof Carlson (George Braziller: New York, 1967).

Page 52.   A clear and authoritative introduction to molecular biology is James Watson's *Molecular Biology of the Gene,* 3rd ed. (W. A. Benjamin, Inc.: Menlo Park, 1976).

Page 55.   The features of cells, the ways that they develop, and the control mechanisms that act within them, are clearly described in *Mechanisms of Development* by Richard Ham and Marilyn Veomatt, (C. V. Mosby Company: St. Louis, 1980).

Page 56.   A classic introduction to Darwinian evolution is George Simpson's *The Meaning of Evolution* (Yale University Press: New Haven, 1950). A somewhat different view of the way large changes in living things arise is given by Niles Eldredge and Stephen Gould in their article, "Punctuated Equilibria: an alternative to phyletic gradualism," published in *Models in Paleobiology,* edited by T. Schopf and J. Thomas (Freeman: San Francisco, 1972).

Page 58.   Gilbert's views about the evolutionary role of introns are described in the article "Why Genes in Pieces?" in *Nature,* vol. 271, p. 501, 1978.

Page 58.   The notion of evolution as tinkering is discussed by François Jacob, in *The Possible and the Actual* (Pantheon Books: New York, 1982).

Page 59.   A discussion of some of the arguments leading to the view of eucaryotic cells as symbiotic associations is given by Lynn Margulis, in *Early Life* (Science Books International: Boston, 1982).

Page 60.   The concept of the biosphere is discussed in the articles of *Scientific American,* vol. 223, no. 3, 1970. See also Chapter 3 of Feinberg and Shapiro's *Life Beyond Earth.* A discussion of the idea that the biosphere is a superorganism is given by J. Lovelock in *Gaia—A New Look at Life on Earth* (Oxford University Press: London, 1979).

Page 62.   The role of deviations from equilibrium and of external sources of energy in maintaining life were discussed by Erwin Schrödinger in his influential book, *What Is Life* (Cambridge University Press: Cambridge, 1944). See also H. Morowitz, *Energy Flow in Biology* (Academic Press: New York, 1968).

Chapter 3

Page 67.   The discoveries of Rutherford, Thomson, et al. are discussed in Chapter 2 of Feinberg *What Is the World Made Of?*, op. cit.

Page 68.   Gunther Stent's views on the future of science are given in his book *Paradoxes of Progress* (W. H. Freeman and Company: San Francisco, 1978). See also his book *The Coming of the Golden Age* (The Natural History Press: Garden City, 1969).

Page 70.   The changes that have taken place in our views of space and time are described in Chapters 4 and 5 of *The Evolution of Scientific Thought*, op. cit.

Page 70.   Some of the properties of black holes are discussed clearly in *Black Holes and Warped Spacetime*, by William Kaufman, III (Bantam Books: New York, 1980). The problems posed for physics by the loss of information in black hole formation have been stressed by John Wheeler, in "Geometrodynamics and the Issue of the Final State," printed in *Relativity, Groups and Topology*, edited by C. DeWitt and B. S. DeWitt (Gordon and Breach: New York, 1964).

Page 76.   The work of Tsung-dao Lee is described by him and Richard Friedberg in their article "Discrete Quantum Mechanics," in *Nuclear Physics*, B vol. 225, pp. 1–52, 1983. This article is more technical than most of the other readings given in this section.

Page 76.   The use of computers to do calculations in quantum field theories is discussed by Michael Creutz in his article "High Energy Physics" in *Physics Today*, vol. 36, pp. 35–42, May 1983.

Page 81.   Some recent work on the dimensionality of space-time is discussed in an article by Edward Witten entitled "Search for a Realistic Kaluza-Klein Theory," published in *Nuclear Physics*, B vol. 186, p. 412, 1981. This article is more technical than most of the other readings.

Page 83.   Some of the issues involved in the directionality of time are clearly discussed by Paul Davies in *The Physics of Time Asymmetry* (University of California Press: Berkeley, 1974).

Page 84.   A discussion of time reversal symmetry, and of some small deviations from it that particle physicists have discovered, is given in Feinberg, *What Is the World Made Of?*, op. cit.

Page 88.   Some of the ideas about the early stages of the universe are discussed in Weinberg, *The First Three Minutes*, op. cit.

Page 92.   A recent discussion of the long-term future of the universe is given by Freeman Dyson, in his article "Time Without End—Physics and Biology in an Open Universe," in *Reviews of Modern Physics*, vol. 51, p. 447, 1979. See also the article by Stephen Frautschi "Entropy in an Expanding Universe," in *Science*, vol. 217, p. 593, 1982.

Page 98.   Physical processes that can take place in a universe that expands indefinitely are discussed by Don Page and M. R. McKee in their article "Matter Annihilation in the Late Universe," in *Physical Review D*, vol. 24, p. 1458, 1981.

Page 99.   The inflationary universe is discussed by Alan Guth in his article "Speculations on the Origin of the Matter, Energy and Entropy of the Universe," published in *Asymptotic Realms of Physics*, edited by A. Guth and K. Juang (M.I.T. Press: Cambridge, 1983).

Page 107.   A learned discourse on the notion of symmetry in science and mathematics is Hermann Weyl's *Symmetry* (Princeton University Press: Princeton, 1952).

Chapter 4

Page 112.   A description of the standard views about the origin of life is given by L. E. Orgel in *The Origin of Life: Molecules and Natural Selection* (John Wiley and Sons: New York, 1973). An interesting alternative is given by A. G. Cairns-Smith in *Genetic Takeover and the Mineral Origins of Life* (Cambridge University Press: Cambridge, 1982). See also, Feinberg and Shapiro, *Life Beyond Earth*, op. cit.

Page 116.   A general overview of what is known about biological development and a discussion of the outstanding problems involved in its understanding is given in Ham and Veomatt, *Mechanisms of Development*, op. cit.

Page 117.   The concept of a flowchart and other ways of representing a program are given in *Fundamental Programming Concepts*, by J. L. Gross and W. S. Brainerd (Harper and Row: New York, 1972).

Page 124.   An interesting alternative approach to morphogenesis is presented by Thom, *Structural Stability and Morphogenesis*, op. cit.

Page 125.   A readable discussion of what is known about aging and some theories of aging is given by B. L. Strehler in *Time, Cells and Aging* (Academic Press: New York, 1962). A more recent summary is in the *Handbook of the Biology of Aging,* by C. Finch and L. Hayflick (Van Nostrand Reinhold Co.: New York, 1977).

Page 130. P. Medawar's evolutionary model of aging is described in the article "Old Age and Natural Death," reprinted in his book *The Uniqueness of the Individual* (Dover Publications: New York, 1981).

Chapter 5

Page 138. A summary of the expected properties of gravitational waves, and some efforts under way to detect them is given by the American physicist Kip Thorne, in his article, "Gravitational Wave Research," *Reviews of Modern Physics,* vol. 52, p. 285, 1980.

Page 147. A description of the DUMAND detector is given in several articles published in the *Proceedings of the 1980 DUMAND Symposium,* edited by V. J. Stegner (Hawaii Dumand Center: 1980).

Page 152. The process of holography, especially as it might be applied to study life processes, is described by J. C. Solem and G. C. Baldwin in their article "Microholography of Living Organisms," *Science,* vol. 218, p. 229, 1982.

Chapter 6

Page 160. The history of the use of mathematics in science is learnedly described by Saloman Bochner in *The Role of Mathematics in the Rise of Science* (Princeton University Press: Princeton, 1981).

Page 164. Thom's views and program are expounded in *Structural Stability and Morphogenesis,* op. cit.

Page 164. D'Arcy Thompson's major work is his book *On Growth and Form* (Cambridge University Press: Cambridge, 1942). A modern work that covers similar subjects is Peter Stevens' *Patterns in Nature* (Atlantic-Little Brown: Boston, 1974).

Page 168. Two readable articles on chaos have appeared in the magazine *Physics Today.* Joseph Ford's article "How Random Is a Coin Toss?" (vol. 36, p. 40, April 1983) concentrates on the unpredictability of the solutions of nonlinear equations. Leo Kadanoff describes possible applications of chaotic behavior to actual physical systems in his article "Roads to Chaos" (vol. 36, p. 46, December 1983).

Page 171. Some generalizations about the way order can emerge in diverse situations are described in Hermann Haken's *Synergetics, an Introduction* (Springer-Verlag: Berlin, 1978).

Page 174. I know of no discussion of twistors suitable for general readers. Those readers with a strong mathematical background can find discussions of current research on twistors in the *Twistor Newsletter*, which is privately distributed by the Mathematical Institute of Oxford University.

Page 178. A discussion of the various views of mathematicians and philosophers about the source of mathematical ideas is given in *The Mathematical Experience*, by Philip Davis and Reuben Hersh (Houghton Mifflin Company: Boston, 1982). Another view is described by Philip Kitchen in *The Nature of Mathematical Knowledge* (Oxford University Press: New York, 1983).

Chapter 7

Page 185. The Monte Carlo method of computation is discussed in the book *Monte Carlo Methods* by J. Hammersley and D. Handscomb (Methuen: London, 1965).

Page 186. Feigenbaum describes his discovery in an article titled "Universal Behavior in Non-Linear Systems," published in *Los Alamos Science*, vol. 1, p. 4, 1980.

Page 187. Some aspects of the use of computers in sequencing biomolecules are discussed by T. R. Gingeras and R. J. Roberts in their article "Steps Toward Computer Analysis of Nucleotide Sequences," published in *Science*, vol. 209, pp. 1322–28, 1980.

Page 191 A computer program for doing such Baconian investigations has already been developed. It is described in an article "Studying Scientific Discovery by Computer Simulation" by G. Bradshaw, P. Langley, and H. Simon, published in *Science*, vol. 222, p. 971, 1983.

Page 193. Some of the ideas involved in parallel processing are described in an article "The Use of Concurrent Processors in Science and Engineering" by Geoffrey Fox and Steve Otto, *Physics Today*, vol. 37, p. 53, May 1984.

Page 193. The Columbia parallel processor approach to calculations of the quark-gluon interaction was devised by the American physicists Norman Christ and Anthony Terrano. It is described in their article "A Very Fast Parallel Processor," published in the *IEEE Transactions on Computers*, vol. C-33, 344, 1984.

Page 194.   A summary of early work on artificial intelligence is given in *Computers and Thought,* by J. Feldman and E. Feigenbaum (McGraw-Hill: New York, 1963). Some of the more recent work is described in the profile of Marvin Minsky by Jeremy Bernstein, reprinted in his book *Science Observed* (Basic Books: New York, 1982).

Chapter 8

Page 201.   Babbage's work is described in *The Analytical Engine,* by Jeremy Bernstein (William Morrow and Co.: New York, 1981). See also *Charles Babbage and His Calculating Engines* by P. and E. Morrison (Dover Publications: New York, 1961).

Page 202.   My thoughts on how to decide about how and whether specific technologies are to be used are given in my earlier books, *The Prometheus Project* (Doubleday: New York, 1969) and in *Consequences of Growth* (Seabury: New York, 1977).

Page 204.   A summary of the properties of various types of collapsed matter is given in *Physics of Dense Matter,* edited by C. Hansen (D. Reidel Publishing Co.: Dodrecht, the Netherlands, 1974).

Page 205.   Feynman's speech is reprinted in the book *Miniaturization,* edited by H. D. Gilbert (Reinhold: New York, 1961). For a recent discussion of similar ideas see K. Eric Drexler's article, "Molecular Engineering," in *Proceedings of the National Academy of Sciences,* vol. 78, p. 5275, 1981.

Page 208.   Some of the techniques presently used in microengineering are described by Alec Broers in his article "High Resolution Systems for Microfabrication," printed in *Physics Today,* vol. 32, p. 38, November 1979.

Page 210.   The possibility of designing proteins for specific functions is discussed by K. Ulmer in his article "Protein Engineering," published in *Science,* vol. 219, p. 666, 1983.

Page 218.   Some thoughts about the effect on society of a technology of aging control are given in Feinberg, *Consequences of Growth,* op. cit. See also Alvin Silverstein's *Conquest of Death* (Macmillan: New York, 1979).

Chapter 9

Page 220.   A description of some of the quantitative changes that took place in science up through 1961 is given by D. de Solla

Price in his book *Science Since Babylon* (Yale University Press: New Haven, 1962). More recent data can be found in the *U. S. Statistical Abstract* (Government Printing Office: Washington, D.C., 1983).

Page 224. The competitive aspects of early molecular biology are aptly described by J. Watson in *The Double Helix* (Atheneum: New York, 1968).

Page 224. One of the earliest letter journals was *Physical Review Letters*, which began publication in 1958. It has remained one of the most highly cited scientific journals, and has had many imitators.

Page 224. One of Galileo's ciphers is described by Patrick Moore in *Watchers of the Skies* (G. P. Putnam's Sons: New York, 1974) at page 158.

Page 226. Data on the funding of science in the United States can be found in the *U. S. Statistical Abstract*, op. cit.

Page 229. Scientific research at the Bell Laboratories is described by Jeremy Bernstein in *Three Degrees above Zero* (Scribners: New York, 1984). Research at the Philips Laboratory is discussed by H. Casimir in his autobiography, *Haphazard Reality* (Harper and Row: New York, 1983).

Page 231. A computer network that is being developed among molecular biologists working on DNA sequencing is described in *Science*, March 30, 1984, p. 1379.

Page 231. The computer network among microchip designers is described by E. A. Feigenbaum and P. McCorduck in their book *The Fifth Generation* (New American Library: New York, 1984.)

Chapter 10

Page 244. Perhaps the strongest advocate of mechanical models was the British physicist, Lord Kelvin. See his *Notes of Lectures on Molecular Dynamics and the Wave Theory of Light* (Baltimore, 1884.) One nineteenth-century scientist's negative reaction to the use of mechanical models is given by the French physicist Pierre Duhem, in his book *The Aim and Structure of Physical Theory* (Atheneum: New York, 1962).

Page 244. A beautiful description of a way of thinking current among theoretical physicists in the early twentieth century, including the electromagnetic world picture is given in R. McCormmach's novel *Night Thoughts of a Classical Physicist* (Harvard University Press: Cambridge, 1982).

Page 245.   Einstein's view of the form that physical theory should take is discussed by him in the book *Albert Einstein: Philosopher-Scientist,* edited by P. Schilpp (The Library of Living Philosophers: Evanston, 1949).

Page 246.   The problems involved in long-range weather forecasting due to the instabilities of the equations involved have been discussed by E. N. Lorenz in his article "Deterministic Non-Linear Flow," published in *Journal of Atmospheric Sciences*, vol. 20, p. 130, 1963.

Page 248.   Some steps toward a future science of order have been described by H. Haken in *Synergetics, an Introduction,* op. cit.

Glossary

Amino Acids   Simple organic compounds containing nitrogen, as well as carbon, hydrogen and oxygen, which are the monomers for the polymer chains in proteins.

Antibody   One of a large number of proteins produced by higher animals, which can react with foreign substances within the body, so as to neutralize them.

Antineutrinos   The antiparticles of neutrinos, distinguished from them by the relative orientation of spin and velocity.

Antiparticle   For every known type of subatomic particle, there exists a type of particle with the same mass but opposite electric charge. The two types related in this way are called each other's antiparticle. The first such antiparticles discovered were electrons and positrons.

Antiquark   One of several types of subatomic particles, the antiparticles of the corresponding quarks. Antiquarks bind with quarks to produce unstable hadrons known as mesons.

Artificial Intelligence   A branch of computer science in which attempts are made to duplicate human intellectual functions, such as language translation, by means of computers.

Baconian Model A model of the way science proceeds, described by Francis Bacon. According to this model, scientists accumulate data without reference to prior hypotheses, and then look for patterns in the data.

Big Crunch If the universe is finite, the present expansion will eventually stop and be replaced by a contraction, which in time will reproduce the conditions of high density and temperature that held at the beginning of the universe. This situation has been called the Big Crunch, in contrast to the Big Bang.

Biosphere The collection of all of the material on the surface of earth that is involved in biological cycles.

Black Body An object that absorbs all radiation of any wavelength that hits it, and which itself emits a definite pattern of radiation, depending on its temperature.

Black Hole An object whose gravity is so strong that anything coming close to it cannot escape. Black holes can be thought of as a new type of matter, which is sometimes produced when ordinary matter becomes compressed.

Broken Symmetry A situation, common in quantum field theory (as well as other branches of physics), in which the observed solutions of a set of equations do not display all of the symmetry of the equations themselves.

Catalyst A chemical substance that enhances the rate of a reaction among other substances, without being consumed in the reaction.

Catastrophe Theory A branch of mathematics used to describe how quantities may change suddenly when some parameter that they depend upon changes slightly.

Cell Differentiation A process through which cells in a multicellular organism take on distinct roles, usually differing in their chemical behavior.

Chaos A type of behavior of physical systems in which the evolution of the system cannot be predicted because of its sensitive dependence on minor changes in the properties of the system.

Chromatin The material in the cell nucleus, composed of nucleic acid and certain proteins in a complex geometrical arrangement.

Coherent Radiation Radiation for which the phase varies in a regular way from point to point in space and time.

Coherent Scattering A scattering process in which all of the atoms in an object act together to produce a large effect in the scattering of a particle or wave.

Collapsed Matter Matter whose density is much higher than that of ordinary matter, either because it contains no electrons or because the electrons are held together more tightly.

Computer Network A number of computer terminals, connected by telephone lines so that people can communicate through them.

Connectivity A description of how points in space are joined to one another, which forms part of the subject matter of the branch of mathematics known as topology.

Cosmic Radiation A stream of subatomic particles, mostly high-energy protons, pervading space and sometimes impinging on the earth.

Cosmology The study of the universe as a whole, especially its origin and development.

Cultured Cells Cells grown outside of an organism, usually in some partly artificial growth medium.

Curvature A property of any mathematical space that measures the extent to which it deviates from a Euclidean, or "flat" space.

Decay The spontaneous change of a particle into other, less massive particles. According to quantum theory, the exact time that an individual particle will decay cannot be predicted.

Deterministic Laws are called deterministic if, when a physical quantity has a precise value at one time, the laws allow for the prediction of the value that the quantity has at a later time. The laws of Newtonian physics are deterministic, while those of quantum physics are not.

Development The process by which a multicellular organism goes through the various stages of its life, beginning as a fertilized egg and ending with death.

Dimensions In any mathematical space, each point can be distinguished from any other point by specifying a fixed number of numerical coordinates. This fixed number is called the dimension of the space. In ordinary physical space, the number is three, while in the space-time of special relativity, the dimension is four, because a time coordinate is specified for each event in addition to the space coordinates.

Discrete A set of mathematical or physical objects is discrete if there are gaps between the objects, such as for the whole numbers or the atoms in a solid body.

Dislocations Places in a crystal where the atoms are not arranged in the same regular pattern as the rest of the crystal.

Electromagnetic World Picture A theory in early twentieth-century physics, which tried to account for all phenomena in terms of electromagnetic fields and their interactions.

Electron The lightest subatomic particle that carries electric charge. Electrons are one of the components of atoms in ordinary matter.

Embedded A space is embedded in a space of higher dimensions if the lower-dimensional space can be thought of as being the surface of some region in the higher space. An example is the two-dimensional space defined as the surface of a sphere in a three-dimensional space.

Embryogenesis The stage of development in which a fertilized egg changes into the partially formed organism called an embryo.

Energy A property of matter, involving the speed of objects, their mass, and their relative position. The total energy of a collection of interacting objects does not change with time.

Energy Density The amount of energy present in each volume of space is known as the energy density. How large this density is over the whole universe determines whether the universe is finite or infinite.

Entropy A physical quantity, measuring the amount of disorder in a system. According to the second law of thermodynamics, the entropy of an isolated system can never decrease.

Enzyme A protein molecule that acts as a catalyst for biochemical reactions.

Equilibrium   For any physical system in a specific environment, the state of equilibrium is a situation in which no spontaneous changes will take place in the properties of the system.

Euclidean Geometry   The familiar geometry studied by the ancient Greeks, corresponding to a space without curvature. Different forms of geometry were discovered by nineteenth-century mathematicians, and are called non-Euclidean.

Eucaryotes   Organisms made of one or more cells, each containing a distinct nucleus as well as other structures, such as mitochondria.

Ferromagnet   A substance, usually containing iron, that can be given a permanent magnetization by aligning its atomic magnets.

Feynman Diagrams   A type of picture used by theoretical physicists. The diagrams portray, in rough terms, how to describe certain subatomic particle processes as combinations of simpler processes. They also allow, via a set of rules, for the calculation of the probabilities for each process.

Field, Classical   A classical field is a mathematical quantity describing a region of space in which forces such as electric or magnetic act on some objects.

Field, Quantum   A mathematical quantity, which is used in quantum theory to describe our deepest understanding of nature. Quantum fields can manifest themselves as subatomic particles, or as regions in space in which particles have properties different from what they would be in empty space.

Flowchart   A pictorial representation of a computer program, in which many steps of the program are grouped into a single element of the picture.

Fossil Particles   Subatomic particles that have survived from the early universe, when they were produced in large numbers.

Gamma Ray   A type of electromagnetic radiation with very short wavelength, emitted in some radioactive decays.

General Relativity Theory   A theory devised by Einstein around 1915, which describes the mutual influence of space-time and matter on each other's properties. It accounts for the phenomenon that we call gravity more accurately than does the former theory of Newton.

Gluon   One of several types of subatomic particle that interact strongly with quarks and with each other. As a result of these interactions, quarks and gluons bind together to form hadrons. Gluons have zero mass, spin of 1 unit, and no electric charge.

Graviton   A hypothetical particle, which would be related to gravity in the same way as the photon is related to electromagnetism.

Gravity Waves   When massive objects undergo accelerated motion, they are believed to emit gravitational fields, which travel through space in wavelike patterns, similar to the electromagnetic waves emitted during the accelerated motion of charged objects. Intensive efforts are under way to detect the gravity waves that are thought to be emitted by some astronomical bodies.

Group Theory   A branch of mathematics used in subatomic and other branches of physics. Its fundamental concept is that of a group, a collection of mathematical objects with a definite set of rules, according to which any two can be combined to give a third. Many different groups have been studied by mathematicians.

Hadrons   A type of subatomic particle, including neutrons and protons, which are readily produced in collisions. Hadrons are combinations of quarks bound tightly together.

Holography   A process using coherent radiation to produce three-dimensional images with a great deal of definition.

Homogeneous   A homogeneous universe is one in which conditions are the same in every region, providing that these regions are large enough. Our universe appears to be homogeneous over regions that are a few million light-years across.

Horizon   An imaginary surface surrounding a black hole, such that nothing from inside the horizon can emerge to the outside universe. For a nonrotating black hole of stellar mass, the horizon is a sphere whose radius is a few kilometers.

Inflationary Universe   A cosmological theory according to which our universe went through a period of extremely rapid expansion very soon after it began. At the end of this period of "inflation," the universe settled into the present, more sedate expansion.

Initial Values   When an equation describes how some physical quantity changes with time, it is usually necessary to specify the value that the quantity has at some initial time, in order to predict its values at later time. The input quantities are called initial values.

Interference   The combining of two beams of coherent radiation, producing a result that differs in intensity from point to point in space.

Internal Symmetry   A mathematical relation between the properties of subatomic particles, or of quantum fields, that is true at a specific point in space-time. Other symmetries, such as those included in special relativity theory, relate properties at different space-time points.

Laser Interferometer   A device used to measure separations of objects very accurately. It uses two laser beams, which are made to interfere with each other, and whose interference pattern depends on the separation being measured.

Lattice   A discrete arrangement of points, in three or more dimensions, with a rule determining which points are connected to each other.

Leptons   A class of subatomic particles, which includes electrons and neutrinos. Leptons have spin of ½ unit, and do not interact with gluons.

Linear Momentum   A quality of particles, proportional to mass and velocity, whose total value remains constant when several particles interact.

Lithography   A method for imprinting a pattern onto a surface by selectively removing some of the material from the surface.

Magnetic Monopole   A hypothetical type of subatomic particle, predicted to exist by some quantum field theories. Magnetic monopoles would act as sources of magnetic field. They are expected to be extremely massive, and also extremely rare in our universe.

Mechanical Model   A description in which a physical system is taken to behave like a collection of objects that move according to

Newton's laws of motion. Mechanical models of light and electromagnetism were commonly used in the nineteenth century.

Metamorphosis   A developmental change of an organism from one form into a radically distinct form, as from a caterpillar into a butterfly.

Microfilaments and Microtubules   Long thin strands of protein within cells, whose actions produce motion and changes of shape of the cells.

Microsensor   A tiny device that can be implanted in some object to monitor the way in which some property of the object varies over time.

Microwaves   A form of electromagnetic radiation, whose wavelength is between that of radio waves and that of visible light.

Mitochondria   Parts of the cytoplasm of most eucaryotic cells, containing their own DNA, and involved in the production and transfer of energy within the cell.

Molecular Engineering   The construction of useful objects, on a scale of construction that is molecular in size.

Monomer   A unit, such as an atom or small molecule, many of which can be bound together in a chain or other array to form a polymer.

Morphogenesis   The process by which structures of definite shape, such as tissues, are built up within an organism, through the growth and motion of cells.

Monte Carlo Method   A way of calculating difficult integrals by evaluating the function to be integrated at a random set of points.

Neutrino   One of a number of types of subatomic particles, characterized by no electric charge, spin of ½ unit, small or zero mass, and a low rate of interaction with other particles.

Neutron   A subatomic particle found in the nuclei of atoms. Neutrons carry no electric charge and, like protons, are made of three quarks. Isolated neutrons are unstable against decay into protons, electrons, and neutrinos.

Neutron Star   A collapsed star, most of which is made of neutrons, at an extremely high density.

Nonlinear Equations   Equations in which the second or higher power of the quantity, whose variation is described by the equation, enters into some term of the equation

Nucleic Acids   One of two types (DNA and RNA) of biological substances that are polymers of simple organic bases, arranged on a backbone of sugar and phosphate molecules. In their biologically active forms, DNA usually contains two chains, twisted into a double helix, whereas RNA involves a single chain.

Order   Some physical systems can take on many different configurations. If the system is found in only one or a small number of its possible configurations, it is said to have a high degree of order.

Parallel Processes   Mutually dependent processes that occur at the same time, in a natural system, such as a cell, or an artificial system, such as a computer.

Parameter   The equations that describe any physical situation contain one or more numerical quantities that may vary from situation to situation. These numerical quantities are called parameters.

Pauli Exclusion Principle   A law implying that no two spin ½ particles of the same type can have identical properties at any one time.

Periodic Motion   Motion that returns over and over again to the same place, such as the revolution of a planet around the sun.

Phase   A property of a wave; it describes where a specific point lies along the cycle of the wave from one crest to another. An alternate meaning of phase is one of several forms that a specific substance can take (H_2O, for example, can occur as steam, liquid water, or ice).

Phase Change   An abrupt change in the properties of some substance, or of space, which takes place when the numerical value of some aspect of its environment changes. An example is the freezing of water when it is cooled below zero degrees Celsius.

Photon   The particles that make up light. Photons have neither mass nor charge, but do carry one unit of spin. Photons are the most prevalent particles in the present universe.

Photosynthesis A process carried out by plants and some other organisms, through which the energy in sunlight is used to produce carbohydrates from water and carbon dioxide.

Plasma A form of matter in which electrons and nuclei are detached from one another, and move independently. Plasmas are found in high-temperature conditions, such as inside of stars.

Platonism A view of the nature of mathematics which holds that mathematical objects, such as numbers, exist in a universe of their own, which can be perceived by mathematicians.

Polymer An aggregation of many individual units, called monomers, bound together, usually chemically, into some definite geometrical form, such as a chain. The monomers may be of one or more different types.

Positron A type of subatomic particle; it is the antiparticle of the electron. Positrons are very rare in the present universe, but were almost as common as electrons in the early universe.

Programmed Aging A theory implying that aging is the result of some instructions contained within the genetic material of an organism.

Proteins Biological substances that are based on polymers of some twenty different amino acids. In their biologically active states, proteins may involve several distinct polymer chains, which coil into complex three-dimensional forms.

Proton A subatomic particle found in the nuclei of atoms. Protons carry positive electric charge, and are made of three quarks tightly bound together.

Quantization A mathematical process by which the equations of some pre-quantum theory are changed into corresponding equations of quantum theory. A common result of quantization is that some property that in the pre-quantum theory could vary continuously, such as spin, now takes on only discrete values.

Quantum Chromodynamics (QCD) A quantum field theory describing the interactions of the fields corresponding to quarks and gluons, and thought to explain the properties of hadrons.

Quantum Electrodynamics (QED) A quantum field theory describing the interactions of the fields corresponding to electrons, or other charged particles, with the electromagnetic field. QED furnishes a very accurate explanation of the properties of atoms.

Quantum Gravity A theory, not yet fully developed, which would take into account the influence of quantum theory on the behavior of gravity.

Quark One of several types of subatomic particle. Two types of quark are the constituents of neutrons and protons, the particles found in the nuclei of ordinary matter. Quarks are thought to occur only bound together in groups, so that the evidence for their existence is indirect.

Quasars Astronomical objects, some very distant from us, which emit large amounts of radiant energy.

Regularization A mathematical procedure through which certain of the equations of quantum field theory, which seem to predict infinite answers for questions of physical interest, are modified so that the answers are finite.

Relativistic Quantum Mechanics A theory which combines the principles of quantum mechanics with those of Einstein's special relativity theory. Quantum field theories are one part of relativistic quantum mechanics.

Rest Energy A form of energy carried by an object even when it is not moving and is far from all other objects. The rest energy of any object is proportional to its mass.

Restriction Enzymes Proteins which cut double-stranded DNA at selective places, at which specific subsequences begin or end.

Ribosomes Parts of a cell, made of protein and a form of RNA. Ribosomes are the sites of protein synthesis within the cell.

Scattering A process in which some property of a particle or wave, such as linear momentum, is changed as a result of its interaction with a target object. For example, when a wave hits a target, the direction of the wave may change as the result of scattering.

Self-Assembly The spontaneous organization of the components of some structure, such as ribosomes, to produce the structure.

Sequencing The determination of the order of the bases along a nucleic acid chain or of the amino acids along a protein chain.

Shotgun Sequencing   A method for sequencing nucleic acids that involves cleaving the initial chain into various overlapping segments, and then reconstructing the sequence from a knowledge of these segments.

Singularity   A region of space-time where the effects of gravity are so great that the usual description of physical phenomena is not applicable.

Space-Time   According to special relativity theory, different observers will measure different values for the space and time coordinates of various objects. Because measurements of space and time become intertwined in this way, it is mathematically attractive to describe physical laws by using four dimensions, three of space and one of time. The resulting mathematical system is called four-dimensional space-time.

Special Relativity Theory   A physical theory, first devised by Einstein, that describes the relation between the way a physical system looks to observers who are moving uniformly with respect to one another.

Spin   An attribute of subatomic particles, analogous to rotation about an internal axis. In quantum theory, the spin of a particle can only take on certain values, which are multiples of a basic unit.

State   In quantum mechanics, when we know all that we can at any time about any physical system, the system is said to be in a state. For example, an electron is in a state when we know its linear momentum and its spin direction.

Strong Interactions   One of several processes that can lead with high probability to a change in the number or type of subatomic particles present in some region.

Subatomic Particles   According to present understanding among physicists, the ultimate constituents of all matter. Subatomic particles carry such properties as energy and charge. Large numbers of such particles compose the smallest visible bits of matter.

Supersymmetry   An internal symmetry of some proposed quantum field theories, which relates fields describing particles with spin ½ to fields describing particles with spin 0 or 1.

Symmetry   When one or more objects have properties that are the same or that are related to one another in a simple way, the related properties are said to display a symmetry. An example is that the energy of an atom in empty space is the same whatever the direction of its spin.

Synchrotron   A type of accelerator for subatomic particles. Synchrotrons that accelerate electrons emit large amounts of electromagnetic radiation, known as synchrotron radiation.

Tensor Analysis   A branch of mathematics used in general relativity theory. Tensor analysis deals with how different observers see various aspects of space, time, and other physical quantities.

Thermodynamics   The branch of physics that deals with properties of bulk matter such as temperature and heat changes.

Time Reversal Symmetry   A principle of subatomic physics, implying that the rate of any physical process is equal to the rate of the process obtained by reversing the direction in which time passes. Some uncommon decay processes do not satisfy this principle, and the reason for this is not completely known.

Topology   A branch of mathematics which deals with such questions as how points in space can be connected, and with those properties of a space that remain unchanged as the space is distorted in various ways.

Transcription   A process by which the information in a DNA strand is used to produce an RNA strand with equivalent information.

Transformed Cells   Cells that have gone through some genetic change that allows them to divide indefinitely.

Turbulent   At high speeds, fluids flow in a haphazard way, in which there is no relation between the direction or speed of flow at nearby points. Such flow is called turbulent flow.

Twistors   Twistors are mathematical objects consisting of pairs of complex numbers. Their use as an alternative way of describing space-time has been proposed.

Vacuum Fluctuations   Changes in the amount of quantum fields that are present in otherwise empty space, which occur again and again, for very short time intervals, as allowed by Heisenberg's uncertainty principle.

Wavelength   A property of a wave, measured by the distance between successive crests. Different colors of light correspond to different electromagnetic wavelengths.

Weak Interactions   One of several processes that can result in a change in the number or type of particles present in some region, but only with low probability.

Wear-and-Tear-Aging   A theory implying that aging is the result of some environmental influences that disturb the normal functioning of the organism.

White Dwarf   The end point of evolution of many stars, in which the atomic nuclei form a tightly bound lattice and the electrons move freely through the lattice. The matter in most white dwarfs is at much higher density than that of ordinary matter.

W-Particle   A subatomic particle, first observed in 1983, involved in a type of interaction called weak interactions.

Zygote   The fertilized egg before it has begun to divide.

Illustration Credits

Figure 3   Reproduced by permission of Encyclopedia Britannica, Inc.; illustration by Richard Roiniotis.

Figure 12   Reproduced by permission of Encyclopedia Britannica, Inc.; illustration by John Youssi.

Figure 13   Reproduced from *Fundamental Programming Concepts,* by permission of Dr. Walter Brainerd and Dr. Jonathan Gross.

Figure 14   Reproduced from a drawing by Catherine Verhulst in the book *Mechanics of Development* by permission of Dr. Richard Ham.

Figure 15   Reproduced from a photograph taken by Dr. David Alcorta, with his permission and that of Dr. Robert Pollack.

Figure 16   Reproduced from *Time, Cells, and Aging* by permission of Dr. Bernard Strehler.

Figure 17   Reproduced from a photograph by Dr. Bruce Stillman, by permission of the Cold Spring Harbor Laboratory.

Figure 18 Reproduced from a photograph of a detector built at Caltech by a group headed by Dr. Ronald Drever, with their permission and that of Dr. Kip Thorne.

Figure 19 Reproduced by permission of Dr. Vince Peterson and Hawaii DUMAND Center.

Figure 22 Reproduced from *Catastrophe Theory* By E. C. Zeeman by permission of M. Schaeffer.

Figure 23 Reproduced from a photo taken at CERN Laboratory, with the permission of CERN and North-Holland Physics Publishing.

Figure 24–25 Reproduced from the journal *Nucleic Acids Research* by permission of IRL Press and Dr. John Harding.

Figure 26 Reproduced by permission of Dr. Norman Christ and Dr. Anthony Terrano; from their article "A Very Fast Parallel Processor," published in the *IEEE Transactions on Computers,* vol. C-33, 334, 1984.

Figure 28 Reproduced by permission of IBM research.

Index